Introduction to Polyphasic Dispersed Systems Theory

Jacques Thierie

Introduction to Polyphasic Dispersed Systems Theory

Application to Open Systems of Microorganisms' Culture

Jacques Thierie
Brussels
Belgium

ISBN 978-3-319-27852-0 ISBN 978-3-319-27853-7 (eBook)
DOI 10.1007/978-3-319-27853-7

Library of Congress Control Number: 2015958331

Printed on acid-free paper

This Springer imprint is published by SpringerNature
The registered company is Springer International Publishing AG Switzerland

To Ilya Prigogine, my former professor, promoter, and research director, who taught me that even in the most rigorous science, freedom to create and to think must remain autonomous.

To my wife, for her abnegation and her patience.

Contents

1 General Introduction 1
References 8

2 General Results 11
2.1 Definition of a Polyphasic Dispersed System (PDS) 11
2.2 Characteristic Volume and Pseudohomogeneity 14
2.3 Reaction Concentration and Pseudohomogeneity 20
2.3.1 Consequences of the Definition of Pseudohomogeneity . . . 21
2.3.2 Phase Density 22
2.3.3 Phase Volume Relationship 23
2.3.4 Relationship Between R- and E-Concentrations 23
2.4 Mass Balances 24
2.4.1 Note Concerning the Variational Term Appearing in (2.4.12) 30
2.5 The Grouping Principle 31
2.6 Comment on a Certain Relativity in Writing the Mass Balance . . . 32
2.6.1 Discussion 34
2.6.2 Example 35
2.7 Influence of Phase Density on Concentrations Calculation 37
2.7.1 Two-Phase System 39
2.7.2 Numerical Example 40
2.7.3 Critical Biomass 41
2.7.4 The Case of a Gaseous Phase 42
2.8 The Variation of the Internal Composition of a Microorganism with the Growth Rate Is a Consequence of the Mass Conservation Law 42
References 45

3 Continuous Culture: The Chemostat . . . 47
3.1 General Remarks . . . 47
3.1.1 Historical Overview . . . 48
3.1.2 The New Paradigm . . . 49
3.2 The Chemostat and Monod's Model . . . 50
3.2.1 The Beginnings . . . 50
3.2.2 Summary Presentation of the Chemostat . . . 51
3.2.3 Monod's Model . . . 53
3.2.4 Stability of Steady States . . . 55
3.3 The Chemostat in PDS Theory . . . 60
3.3.1 The General Mass Balance . . . 60
3.3.2 The Biphasic Chemostat . . . 62
3.3.3 Steady States in Simple Situations . . . 64
3.3.4 Partial, Specific Rates . . . 64
3.3.5 The Water Problem . . . 72
3.3.6 Examples . . . 73
3.4 The Concept of Maintenance . . . 78
3.4.1 Introduction . . . 79
3.4.2 Method . . . 81
3.4.3 Results . . . 82
3.4.4 Substrate Recycling . . . 84
3.4.5 Practical Considerations . . . 87
3.4.6 Discussion . . . 89
3.4.7 Additional Considerations . . . 91
3.4.8 Should the Concept of Maintenance Be Abandoned? . . . 93
3.5 Threshold Phenomena, Signals in Cells, and Metabolic Pathway Switches . . . 94
3.5.1 Notice . . . 94
3.5.2 Summary . . . 95
3.5.3 Model with One Pathway $n = 1$ . . . 95
3.5.4 Model with Two Pathways $n = 2$ . . . 96
3.5.5 Introduction . . . 97
3.5.6 Implicit Mass Balance . . . 98
3.5.7 Explanation of Specific Rate . . . 101
3.5.8 Discussion . . . 120
3.5.9 Remarks on the Coupling of Maintenance Energy and the Metabolic Switch . . . 123
3.6 The Crabtree Effect in *Saccharomyces cerevisiae* . . . 124
3.6.1 Introduction . . . 124
3.6.2 Result . . . 131
3.6.3 Discussion . . . 153
3.7 Respiro-Fermentative Phenomena in Bacterial Flocs . . . 156
3.7.1 Introduction . . . 156
3.7.2 Materials and Methods . . . 157

3.7.3 Results . . . 158
3.7.4 Modeling . . . 163
3.7.5 Estimation of Subtrate in the Medium . . . 166
3.7.6 Discussion . . . 168
References . . . 170

4 General Discussion . . . 175
4.1 About the Theory . . . 175
4.2 On the Subject of the Applications . . . 178
References . . . 184

Appendix A.1: Theory . . . 187

Appendix A.2: Definitions . . . 191

Appendix A.3: Specific Growth Rate for the Whole . . . 197

Appendix A.4: Cancelling of Interphasic Exchange Fluxes at the System Level . . . 201

Appendix A.5: Consideration about the Micellian Variational Term . . . 203

Chapter 1
General Introduction

Le refus positif d'égarer la discussion dans les espaces vides de l'abstraction se traduit chez le biologiste par un engagement extrêmement méthodique dans la voie de l'expérimentation, et cela très généralement sous la forme de l'expérimentation causale. On s'efforce de démonter un mécanisme, de ramener son fonctionnement à un schéma d'où se trouve –très justement – exclue toute référence à un système de concepts abstraits préconçus. Mais cette précipitation à expliquer, à triompher d'un problème – qui est très généralement un problème d'infime détail – exclut une vertu épistémologique qu'il faut bien retenir comme fondamentale, la disponibilité théorétique.

François Meyer, (in 1967; Ref. 1986).

The biologist's positive refusal to get entangled in the undefined space of abstraction leads to a very systematic commitment to experimentation, in most cases in the form of causal experiments. He strives to dismantle a mechanism, to sum up its functioning in a pattern excluding – and rightly so – any reference to a preconceived system of abstract notions. However, such a hasty endeavour to explain, overcome a problem – typically only a matter of insignificant detail – results in the exclusion of a pivotal epistemological principle, that of the theoretic opportunity.

Translation according to Meyer. (Thanks to Prof. M.-F. Boeve, I am grateful for her help with the English translation.)

In their introduction to the book, "*The Biotechnology Revolution?*" (Fransman et al. 1995), the authors place biotechnology in the "new technological paradigm," along with microelectronics, IT, and new materials. In this particular context (that of the new paradigm), biotechnology must obviously be made up of the same parts as new (or modern) biotechnology (just as it is for example, defined socioeconomically by Freeman 1995). It is not made up of the group of separate (bio)techniques, essentially fermentation, that have accompanied humanity for centuries, even for millennia, and which have given us alcohols, cheeses, sauerkraut, and so many other foods.

Whether modern biotechnology is a real technological revolution, or not, is however beside the point. What attracts us more from the incontestable rise of the

J. Thierie, *Introduction to Polyphasic Dispersed Systems Theory*,
DOI 10.1007/978-3-319-27853-7_1

phenomenon is the evidence, within the domain of ideas, of the status of life sciences. Both socially and economically, considerable advances in medicine, food-processing, and in environmental studies obviously still conferred an unimaginable importance to biology up to the nineteenth century and even up to the beginning of the twentieth century. However, this does not constitute a change in the paradigm (in the sense of "disciplinary matrix" from Kuhn (Chalmers 1988; Kuhn 1983)). The intervening modification of nature in the science of living things is linked to its rise to the rank of technology. Basically, biotechnology associates living things and the process of production that has been mastered (for example). Thus appeared a pluridisciplinary approach that integrated materials that were essentially biological (biochemistry, microbiology…) and techniques that could be qualified as physicochemical. These techniques display features that are the fruit of a long evolution and they are both theoretical and quantitative. In reality, it is often the theorization of a group of phenomena that makes them quantifiable. So, biotechnology appears to us as the emergence of a pluridisciplinary material where the constraints on technical and economical performances are going to extend the properties of quantification and theorization of the "hard sciences" to the description of living things.

Of course, the rough outline of such a step preceded the demands of modern biotechnology. As Chalmers (1988) pointed out, "From the Kuhn's point of view, the type of factors that contribute to the facts that cause changes of the scientific paradigm is a subject of psychological and sociological research." Without wanting to make a historical record of this attempt, we can perhaps bring back to mind a notable event that ended in 1937 when Ludwig Von Bertalanffy presented his "general theory of systems" to the University of Chicago. He relates, "Unfortunately, at this time, the biological theories were not well received and I was frightened by the 'clamor of the Beotians' as the mathematician Gauss used to say. I left my broths in a drawer and it was only after the war that my first publications on the subject appeared (Von Bertalanffy 1973). Some forty years later, the Société française de biologie théorique, (French Society of Biological Theory) that held a "seminar [proposing] reflection on the basis and justification of discourse explaining living things" (Lück 1985). So there was from (let's say) the failure of Von Bertalanffy to the present day, a considerable development in the attitudes of biologists (but not of all biologists) to an attempt to theorize their field of knowledge. At present, the reduction of complex systems to formulae (Nicolis and Prigogine 1989) throughout the physics of nonequilibrium and nonlinear dynamics opened a considerable heuristic space to biologists in a form that could be defined (in a broad sense) as mathematical modeling (Brown and Rothery 1993; Haefner 1996; Kauffman 1993; Murray 1993, and many others). The themes covered by this type of approach are very diverse and extend from the analysis of intracellular phenomena to the dynamics of populations, by way of neuronal systems. Nevertheless, we could qualify this thinking as very heterogeneous, and as stemming from the essentially academic. The diversity of not only the methods but also the objectives (description, understanding, classification) prevents us from considering this group of techniques of representation as a unique and structured discipline. So, in this way, a whole scientific community foresaw the "theoretical availability" of biology that François Meyer evoked in the text in question, but from our point of view, without attaining it and without even approaching the epistemological situation of physics that is so often taken as a model when this sort of problem is approached. Perhaps we should admit that a biological theory is impossible because of its object of study, or that there are misgivings over criteria of truth that can differ from one science to another (Buscaglia et al. 1983), but that is another story.

It was necessary to make this wide digression to pay homage to a basic step that has so enriched biology, and especially to differentiate between two current theories that developed just about simultaneously, sometimes using the same tools, but which remained in parallel (if not unknown to each other) probably because they did not concern themselves with the same problem and doubtless because they did not have the same objectives. The second theoretical approach of life sciences falls within the scope of medical, environmental, and above all, industrial biotechnology and is therefore subject to the constraint of the quantification of results (that serves to optimize the processes of production both technically and economically). Quantifying and theorizing are certainly not identical steps, but it is very possible that the effort of quantifying ends up by putting a stop to the production of theoretical tools which, when they interact could give rise to what we are obliged to call a "theory." Our objective is to try to place the step begun in this work in a more general historical and conceptual perspective. I am neither a historian nor a philosopher so this attempt should be taken as a pedagogical attempt to restore the work developed in the following chapters to a more general field, namely that of post-genomic metabolic engineering (ME) (Kholodenko and Westerhoff 2004) and perhaps even beyond.

Within the scope of microbiology, which is our field of application, the study of the growth of microorganisms was an integrated domain for an approach to the phenomenon through several disciplines. Although an approach to kinetic chemistry appeared very early (Hinshelwood 1946; Nobel Prize for Chemistry) "with the objective of sensitising chemists to the fact that microbiology presents many problems that concern them" it seems to us that the real initial rise toward understanding growth was by its very nature essentially physiological, perhaps from the original impetus of the Copenhagen School of de Maaløe (Maaløe and Kjelgaard 1966), that became progressively wider while keeping a certain unity of thought right up to the present (Ingraham et al. 1983; Neidhardt et al. 1990). This approach that we class as physiological has had great success and the principal result of its research has doubtless been to show that the composition of a bacteria (at least at macromolecular level) depended on its speed of growth and not on the milieu of growth. It was perhaps in 1975 (Kjelgaard and Maaløe 1976) that this current physiology was going to gather strength. During a symposium in 1975, in the chapter, "Retrospectives—Perspectives," von Meyenburg expressed the feeling that many facts had been collected without anyone being able to relate them. Neidhardt replied that the new genetics and biochemical tools seemed to be what was required to speedup progress (tools for the understanding of monitoring synthesis of from ribosomes). At this stage, it appears that genetics and biochemistry (quite rightly) cannot be overlooked attempting to understand the cell group phenomenon, i.e., growth. However, Neidhardt realized the danger of losing the integration aspect of research when applying himself to an approach that was too reductionist. He writes, "*Looking ahead, we can see that we shall want to integrate numerical information about not only all the elements just mentioned* [...] *but also the same information about the rest of the cell. This integration is an absolute requirement for the final test of any hypothesis of cell control, whether based on* in vitro *or genetic evidence, or both. Unfortunately, the present pattern of research does not inspire confidence that this integration of data will be possible.*" We think that this tirade from Neidhart was prophetic and that effectively the end of the twentieth century was largely dominated by disciplines and approaches that are reductionist. This stage surely was (and remains) necessary, but it is definitely time to pay more attention now to an effort to integrate the facts. In 1990, Sir James Black, Nobel Prize 1988 for Physiology and Medicine, declared to the British press in a provocative manner, that the future of pharmacology would be "the progressive triumph of physiology over molecular biology." Many physiologists, with different specialist fields, have stood by this declaration and their thoughts have been published in 1993 (Boyd and Noble 1993). This publication resounds something like an echo of Neidhardt's anxiety and shows that in many disciplines linked to life sciences an effort toward the integration step is very desirable, if not required.

From the microbiological point of view and from the point of view of its industrial applications the rise in classic biotechnologies can be placed around 1940 and the rise in new biotechnologies around 1970—as defined by their new fields of study and their genetic engineering techniques. The influence of chemistry has been known since the beginning of the twentieth century (the first handbook was revised by George E. Davis en 1901) and his principles are obviously applied to biotechnologies (Schügerl 1987, 1991, 1997; Schügerl and Bellgaardt 2000; Johnson 1999). Two remarks, however, remain to be made at this level. First, if the technique of continuous reactors is well known to chemistry (Villermaux 1982), it was discovered in 1949 by Monod, using bacterial cultures (refer to the chemostat Sect. 3.2) and was not established at an industrial level (except for the production of beer in continuous and this appeared in 1968 but has not become very widespread). Second, in efforts to optimize production, improvement in the size of the reactor is often neglected compared with improvement in the cell strain (Stephanopoulos et al. 1998). So it is not surprising to note that the establishment of new biotechnologies around 1970 was accompanied by a series of systems of innovative representation that tried to give a quantitative set of formulae to cell phenomena.

Chronologically, the first approach is perhaps that of Savageau (1969), who published first a "*Biochemical Systems Analysis*" which was to become later the "*Biochemical Systems Theory*" (BST; Savageau 1991). In this latest article, Savageau shows also the relationships between various formulae that derive, according to him, from the BST and he quotes the "*Generalized-Mass-Action*" (GMA), the "*Flux oriented theory*" (FOT), and the "*Metabolic Control Theory*"

(MCT), as well as the representation in "*S-systems*" (that is presented nicely in Voit's book (Voit 1991)).

Next comes the "*Metabolic Control Analysis*" (MCA), written in 1968 but referenced classically by the article by Kacser and Burns in 1973. A good presentation of the theory can be found in Fell (1997).

Torres and Voit (2002) define the "*Metabolic Flux Analysis*" (MFA) at the time the experimental fluxes are determined, by using a mathematical analysis of distributions of these fluxes. (The approach was first used in perhaps 1979 in the Aiba and Matsuoka article (1992), where a stoichiometric method was used for these methods of flux analysis. This approach probably began in 1979 with the article by Aiba and Matsuoka.)

The list of these new approaches is doubtless not exhaustive, but they all appeared around the 1970s, which was an era that saw the rise in modern biotechnologies that (apart from the aspects announced by Freeman 1995) were for the most part integrated with genetic engineering. A group of experimental techniques and mathematical methods of quantification came about simultaneously in this way and they had a common objective that Jay Bailey defined as ME in 1991. In spite of Bailey's critical ("*At present, metabolic engineering is more a collection of examples than a codified science.*" Bailey 1991), the term *metabolic engineering* rapidly brought together numerous researchers or groups of researchers. From 1995, the "*Metabolic Engineering Working Group*" (MEWG) was created following a report from the "*Biotechnology Research Subcommittee*" (BRS) and this was supported by eight federal American agencies (from agriculture to space, by way of defense—refer to "www.metabolicengineering.gov"). In 1998, an important handbook appeared, entitled, "*Metabolic Engineering*," published by MIT and Lingby (University of Denmark), and overseen by Stephanopoulos et al. (1998) and in 1999 appeared the first edition of the review "*Metabolic Engineering*."

At this stage of its development, ME although using intensely biochemistry and genetics, ceased to be strictly reductionist in the sense that it presented an "important and new aspect [that highlighted] an integrated metabolic pathway rather than isolated reactions. In this way, ME takes into account a network of biochemical reactions [...] (Stephanopoulos 1999). In this very same article, Stephanopoulos defines the essence of ME as the coupling of analytical methods (in a mathematical sense) of quantification of fluxes and their control via appropriate genetic modifications. It is this coupling that, under pressure from biotechnologies, will be the driving force of the development of "in silico" methods (in the silicium of computers) that takes control of the bioinformation software and mathematical modeling. Of course the in vivo methods remain to this day.

However, quite quickly the integration of phenomena at the level of metabolic pathways appeared and this was thought inadequate by certain people. In 2002, Cortassa et al. put forward a more holistic version called metabolic cell engineering (MCE) (Cortassa et al. 2002). The accent was put on the necessity of understanding the regulation of metabolic reactions at the cell level and not just at the metabolic pathway level. (A notion that had already occurred to the experimentalists had been brought to the limelight already several years before, (refer to Rossel et al. 2002)).

Kholodenko and Westerhoff (2004) began again a stage in the progression toward a more holistic approach of ME. Not only did they admit that the whole cell must be taken into account, but also the environment. Moreover, in this ME what they qualify as post-genomic, they conceive as a circular causality and not as a dictatorial authority of the genetic code. So they define ME as "*the science-based alteration of the molecules and conditions in and around the living organism with the aim of optimizing both its metabolic productivity and its functioning. The alterations should take account of the response of the organism through its own democratic regulation hierarchy to changes brought about the engineer. It should reckon with the likelihood that improvement of the productivity of the organism should require that homeostasis of the functions that are most important for the organism itself, be maintained.*"

So, at the very beginning of the twenty-first century, we saw emerging a new field called "theoretico-experimental" in life sciences. Its final objective is purely technical and economical since it is essentially a question of producing molecules (metabolites proteins, etc.), that is, at the present time. This should be done as cheaply as possible by procedures that are part of the field of long-term development because a living organism can produce them today … . The first reductionist approaches using genetic engineering were successful but also had resounding failures and involved many failed attempts that failed because the biocomplexity of systems was rejected. Analysis of failure frequency seems to dictate leaving the reductionist approach and tending toward a more holistic step. Now, someone announced one day a curious principle of incertitude that could be put in the form of equations in this way (I regret that I cannot acknowledge its author since I have lost the reference but I hope not to betray his line of thought):

$$\text{generalization} \times \text{precision} = \text{constant} \tag{1.1}$$

This can be put in the terms: *the more widespread a system of description is, the less the degree of precision with which it can be described is refined.* Ignoring this principle is precisely the challenge that is presented by ME, that is, to (and it is no joke) integrate a whole organism in the Universe in order to extract one tiny molecule.

The solution we have come up with in an effort to ignore the principle of incertitude (1.1) was to consider the whole of the system to be studied, in the form of subsystems. The more a subsystem is advanced, the less widespread it is and the easier it is to define it with precision. In concrete terms, the situation is presented as in Fig. 1.1. The stage of a subsystem (in Roman numerals) advances from top to bottom.

The system to be described (at the level of the whole system: bioreactor—stage 1) is included in the exterior environment (the Universe—stage 0) with which it exchanges material and energy. It is therefore an open system and the exchange between the two stages is vertical or at the entry or exit. One of the subsystems (at the level of phase II.1—stage II) is formed, by example, from cells of types **a**, whereas another (at the level of phase II.3—stage II) is formed from the nutritive milieu. So there is an exchange of material and/or energy between these sublevels. These

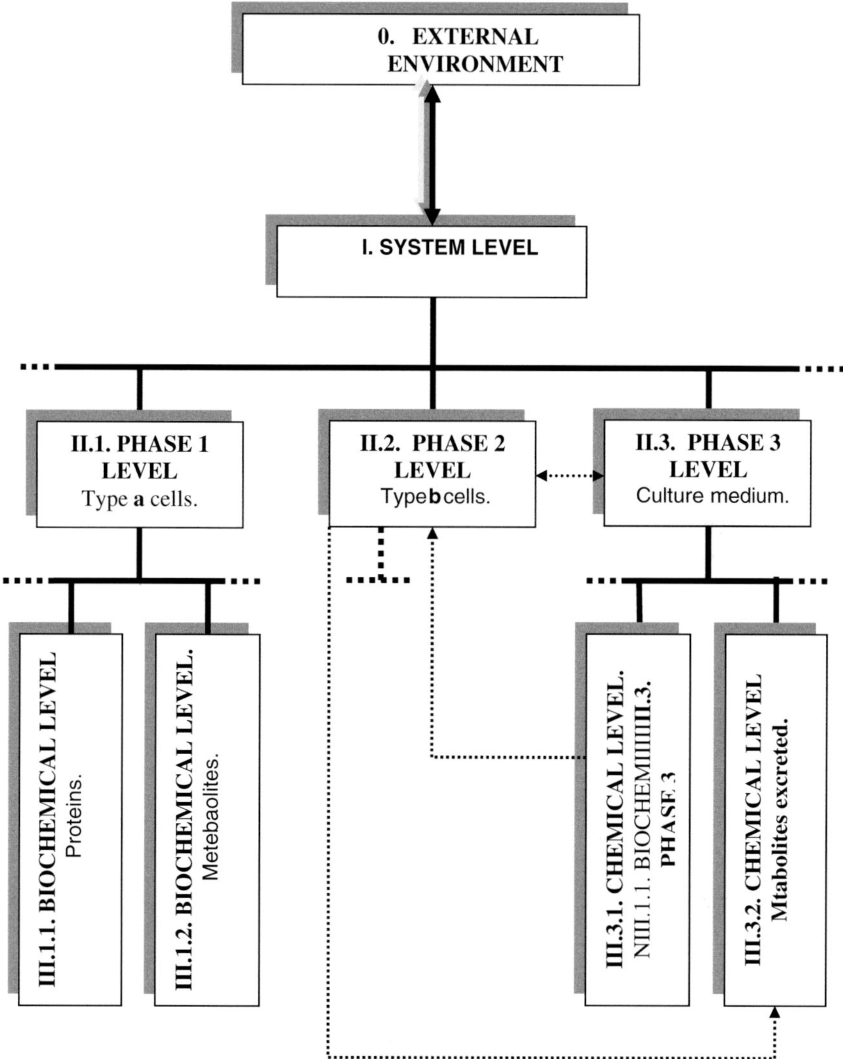

Fig. 1.1 Representation of the sub-levels of description in SPDs. Four stages of levels of description are represented (from 0 to III). The diagram shows a hierarchical structure moving from the most general and the most 'vague' level (the Universe) towards the most restricted and precise level (bio) chemical level. The dashed arrows show the most important exchange fluxes (with the exception of input/output fluxes). Refer to the text for more details

exchanges of material and/or energy are called horizontal or interphasic. Each level of phase can still be subdivided into sublevels of stage III, which are here molecular levels and which can be understood in great detail (Sect. 2.5). The grouping principle makes possible the reduction or increase in the number of these levels in accordance with the amount of detail desired for the description.

We have not developed the intermediary level that includes cell organites (or organelles) because it is felt that there is no need for this with the applications that we have developed. (Such a representation would constitute a polyphasic dispersed system that was compartmentalized.)

Finally, we have chosen the chemostat as bioreactor because it is a system in which all materials can be used (and not just gases as in the case of batch, for example) and because it is subject to important constraints such as the external environment via entry flux. At the systemic level, it represents an ideal situation. However, the chemostat also represents a bioreactor that is just about ideal for studying the properties of cells from a physiological point of view. This fact is recognized by a growing number of authors and is made quite clear by Panikov (1995).

To this must be added that the chemostat constitutes the archetype of the continuous culture of microorganisms and cells. Theoretically, once its optimum stationary state has been reached, a production of biomass or molecular components could be maintained indefinitely. This situation is enormously advantageous for numerous industrial processes (if not all of them). However, this method of production remains extremely marginal in the industry. This state of affairs will not be discussed here, but it is thought that there are as many psychological blocks as objective reasons attached to this situation. It is hoped that in addition to its theoretical aspect, the dispersed polyphasic systems approach (non-compartmentalized systems) that is presented in the pages that follow will help also to implement continuous biotechnical procedures that favor long-term development and socioeconomic growth that is devoted to benefiting everyone.

References

Aiba S and Matsuoka M (1979) Identification of metabolic model: citrate production from glucose by *Candida lypolitica*. Biotechnol. Bioeng. **21**:1373–1386.

Bailey JE (1991) Towards a science of metabolic engineering. Science. **252**: 1668–1674.

Boyd CAR and Noble D (Eds.) (1993) The logic of life – The challenge of integrative physiology. Oxford University Press, New York.

Brown D and Rothery P (1993) Models in biology: Mathematics, statistics and computing. John Wiley and Sons Ltd, Chichester, England.

Buscaglia M, Lalive D'Epinay C, Morel B, Ruegg H et Vonèche J (Eds) (1983) Les critères de vérité dans la recherché scientifique – Un dialogue multidisciplinaire. Actes du Colloque de l'Association des Professeurs de l'Université de Genève. Maloine SA Editeur, Paris.

Chalmers AF (1988) Qu'est-ce que la science? Editions La Découverte, Paris.

Cortassa S, Aon MA, Iglesias AA and Lloyd D (2002) And introduction to cellular and metabolic engineering. World Scientific C°, Singapore.

Fell D (1997) Understanding the control of metabolism. Portland Press Ltd., London, UK.

Fransman M, Junne G and Roobeek A (Eds) (1995) The biotechnology revolution? Blackwell Publishers, Oxford, UK.

Freeman C (1995) Technological revolutions: historical analogies. In "The biotechnology revolution?" Blackwell Publishers, Oxford, UK.

Haefner JW (1996) Modeling biological systems – Principles and applications. Chapman & Hall, New York, USA.

Hinshelwood CN (1946) The chemical kinetics of the bacterial cell. Oxford University Press, Clavendon.

Ingraham JL, Maaløe O and Neidhardt FC (1983) Growth of the bacterial cell. Sinauer Associates Inc., Sunderland, Massachussetts.

Johnson AT (1999) Biological process engineering. – An analogical approach to fluid flow, heat transfer and mass transfer applied to biological systems. John Wiley & Sons Inc., USA.

Kacser H and Burns JA (1973) The control of flux. Symp. Soc. Exp. Biol. **27**:65–104.

Kauffman SA (1993) The origin of order – Self-organization and selection in evolution. Oxford University Press, New York.

Kholodenko BN and Westerhoff HV (2004) Metabolic engineering in the post-genomic era. Horizon Bioscience, Wymondham, UK.

Kjelgaard NO and Maaløe O (1976) Control of ribosomes synthesis. Proceedings of the Symposium – Copenhague 2–5 June 75. Academic Press, New York.

Kuhn TS (1983) La structure des révolutions scientifiques. Trad. Laure Mayer. Flammarion, Paris.

Lück HB (Ed) (1985) Biologie théorique – Solignac 1985 Editions du CNRS, Paris.

Maaløe O and Kjelgaard NO (1966) Control of macromolecular synthesis. A study of DNA, RNA and protein synthesis in bacteria. WA Benjamin Inc, New York.

Meyr F (1986) Situation épistémologique de la biologie. In "Logique et connaissance scientifique (Dir. J Piaget) Collection La Pléiade, Gallimard.

Murray JD (1993) Mathematical biology. Biomathematics. Vol. 19. Springer-Verlag, Berlin, Heidelberg.

Neidhardt FC, Ingraham JL and Schaechter M (1990) Physiology of the bacterial cell. A molecular approach. Sinauer Associates Inc., Sunderland, Massachussetts.

Nicolis G and Prigogine I (1989) Exploring complexity – An introduction. WH Freeman and C°, New York, USA.

Panikov NS (1995) Microbial growth kinetics. Chapman & Hall, London, UK.

Rossel S, Coen C, van der Weijden C, Kruckeberg A, Bakker BM and Westerhoff HV (2002) Loss of fermentative capacity in baker's yeast can partly be explained by reduced glucose uptake capacity. Molecular Biology Reports. **29**: 255–257.

Savageau MA (1969) Biochemical system analysis. I. Some mathematical properties of the rate law for the components enzymatic reactions. J. Theor. Biol., **25**:365–369.

Savageau MA (1991) Biochemical system theory: operational difference among variant representations and their significance. J. Theor. Biol., **151**: 509–530.

Schügerl K (1987) Bioreaction engineering – Reactions involving microorganisms and cells. Volume 1. Fundamuntals, thermodynamics, formal kinetics, idealized reactors types and operations modes. John Wiley & Sons Ltd., Chichester, UK.

Schügerl K (1991) Bioreaction engineering. Volume 2. Characteristic features of bioreactors. John Wiley & Sons Ltd., Chichester,UK.

Schügerl K (1997) Bioreaction engineering. Volume 3. Bioprocess monitoring. John Wiley & Sons Ltd., Chichester, UK.

Schügerl K and Bellgaardt KH (2000) Bioreaction engineering – Modeling and control. Springer-Verlag, Berlin, Heidelberg, Germany.

Stephanopoulos GN (1999) Metabolic fluxes and metabolic engineering. Metabolic Engineering. **1**:1–11.

Stephanopoulos GN, Aristidou AA and Nielsen J (1998) Metabolic Engineering – Principles and methodologies. Academic Press, USA.

Torres NV and Voit EO (2002) Pathway analysis and optimization in metabolic engineering. Cambridge University Press, UK.

Villermaux J (1982) Génie de la reaction chimique. – Conception et fonctionnement des réacteurs. Technique et Documentation (Lavoisier), Paris.

Voit EO (Ed) (1991) Canonical nonlinear modeling. S-System approach to understand complexity. Van Nostrand Reinhold, New York.

Von Bertalanffy L (1973) Théorie générale des systèmes. Trad. George Braziller. Dunod, Paris.

Chapter 2
General Results

Abstract This chapter is the most general of this book. It does not deal with biotechnology or microbiology in particular. However, it is necessary for these two disciplines which make up the core of this book. In fact, this chapter attempts to find the general theoretical bases of the original approach of polyphasic dispersed systems (PDS). It critically reconsiders the most common assumptions directly used in physical chemistry systems and, in particular, the usual ways of theoretical representation, such as mathematical modeling, for example. Based on the concept of description level described in the previous chapter, it analyzes key hypotheses such as the homogeneity of a system. It attempts, maybe sometimes in a too rudimentary way, to define new concepts, such as the minimum volume of a system required to implement the original concepts of the PDS; it formally introduces the different types of concentrations relevant to a correct description of a phenomenon and the consequences of these definitions on the assessment of experimental measurements. This chapter also provides the basis for calculating the mass balance in a polyphasic dispersed system and shows how this approach can bring to light mass fluxes that are cryptic in the classic modes of descriptions. Finally, last but not least, it draws the attention to the possibility, regarding doubly open systems (in a thermodynamic sense), to envisage a "physicochemical relativity" in writing the mass balances.

2.1 Definition of a Polyphasic Dispersed System (PDS)

Polyphasic dispersed systems such as we consider, differ markedly from the dispersed systems of the physicist. It is principally a two-phase system that belongs to this category, such as aerosols, mousses, emulsions, sols and gels, etc. The dispersed particles are particles of between 10^{-3} and 1 μm and are made up of micelles, bilayers, and bilayer vesicles (of the order of 5–10 nm) (Takeo 1999). In general, these systems display a certain amount of stability in the sense that they can support themselves, over a greater or shorter timescale, even if they are a closed

J. Thierie, *Introduction to Polyphasic Dispersed Systems Theory*,
DOI 10.1007/978-3-319-27853-7_2

system. This property is due to Brownian movement or repulsive forces (electrostatic, for example) according to the size of the particles. They can be the site of transformation processes (aggregation, flocculation, etc.) that can give rise to the separation of phases, but what distinguishes them most, perhaps, from the PDS such as we imagine them, are the dispersion forces. In the cases considered above, these forces are generated by the system itself (essentially repulsive forces; refer to Takeo 1999) and can be considered as internal. In the PDS, on the other hand the dispersive forces are mainly external (such as mechanical agitation, for example). The second fundamental difference concerns the size of the dispersed particles. The scale of the dimensions is between micrometer and millimeter (even centimeter). In spite of major differences, there are obviously things in common that will not be developed here. (These are concepts that have already been brought to mind such as certain repulsive forces (or attraction forces) or even thermodynamic considerations, etc.). So we can almost consider that the class of systems that will be taken into account below is independent of the dispersed systems as presented at the beginning of the section. However, it was important to make clear the distinction to avoid misunderstandings.

A polyphasic dispersed system will be defined as a system composed of several phases (solid, liquid, gaseous) that are closely distributed one in another and for the most part generally maintained in this dispersed state by external dispersion forces such as mechanical agitation. The PDS are thus intrinsically unstable systems. The interface between whatever two phases is therefore discontinuous and limited by phase fragments that we will designate as micelles (to avoid either creating a neologism or having to use periphrases).

Very generally, Fig. 2.1 shows a PDS of n phases $(\phi_1, \phi_2, \phi_3, \ldots \phi_n, \ldots)$.

This PDS is called "compact" because the micelles $(m(\varphi_i))$ form a precise division of the system and it is not possible to travel through a whole phase without crossing at least one interface.

Figure 2.2 shows a dilute PDS where there is a path (C) that makes it possible to travel through the whole phase φ_1 without crossing a single interface. In this case, phase φ_1 is called the "dispersing phase" and it does not form micelles.

There is an intermediary case between the compact and the dilute system; this is called the dense system where dispersing phase micelle can be found (a dispersing phase micelle is a fragment from dispersing phase that is isolated from the phase by micelles from other phases). A phase from a PDS can also be defined as a whole of micelles of the same kind, whether these are in a closed system or an open system and if the input/output fluxes include micelles. (This concept of addition of input/output fluxes will become clearer later in the text.)

In the most general case, there can exist several types of different micelles in one and the same phase. So, for example, the solid phase can be made up of several types of different solid micelles. The liquid phase can also be made up of different liquid micelles formed of non-miscible liquids. (An example in a gaseous phase is not easy to bring to mind.)

Generally, the number of micelles in a PDS (closed or open) is not constant. For example, the number of air bubbles can diminish by coalescence; the number of

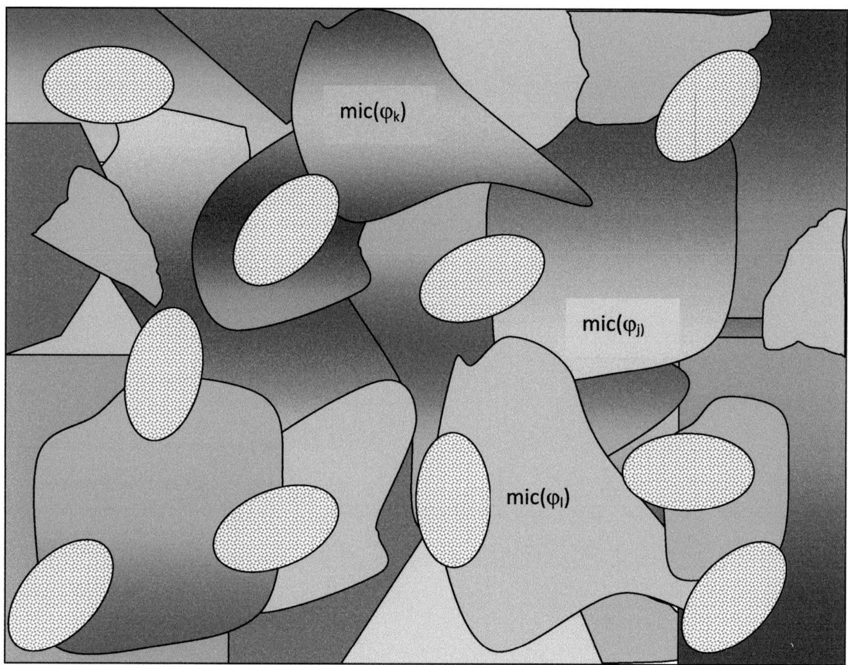

Fig. 2.1 Compact Polyphasic System. A compact PDS gives rise to a perfect division of the system and there is no path along which to travel through an entire phase without crossing at least one interface (mic(φ_i) should be read as "micelle belonging to phase *i*")

solid micelles can diminish by aggregation or increase by fragmentation; two types of micelles can combine to form a new type of micelle; in an open system, inflows and/or outflows of micelles can exist, etc. Nevertheless, if there is a steady state, this is characterized by a number of constant miracles. The steady state can apply to the whole of the micelles in the system or just to certain classes of micelles. By way of an example, an aerated culture can be composed of a constant number of solid micelles (cells, for example), whereas the quantity of gaseous micelles (bubbles) can vary over time.

However, the preceding considerations are not enough to define completely the PDS. In the way it has been described until now, the PDS appears clearly as a heterogeneous system. To complete the definition, the notion of pseudohomogeneity must be introduced. This means that in spite of the intrinsic heterogeneous character of the system, there is a certain level of description in which the element of the volume of the system can be considered as homogeneous (from a functional point of view). The element of volume in question is necessarily much bigger than the mean volume of one micelle but much less than the volume of the entire system. If this notion of pseudohomogeneity does not apply, the whole of the micelles and phases, etc., that are described above would only constitute a collection of objects that could not be treated as a continuous medium. Now this property is

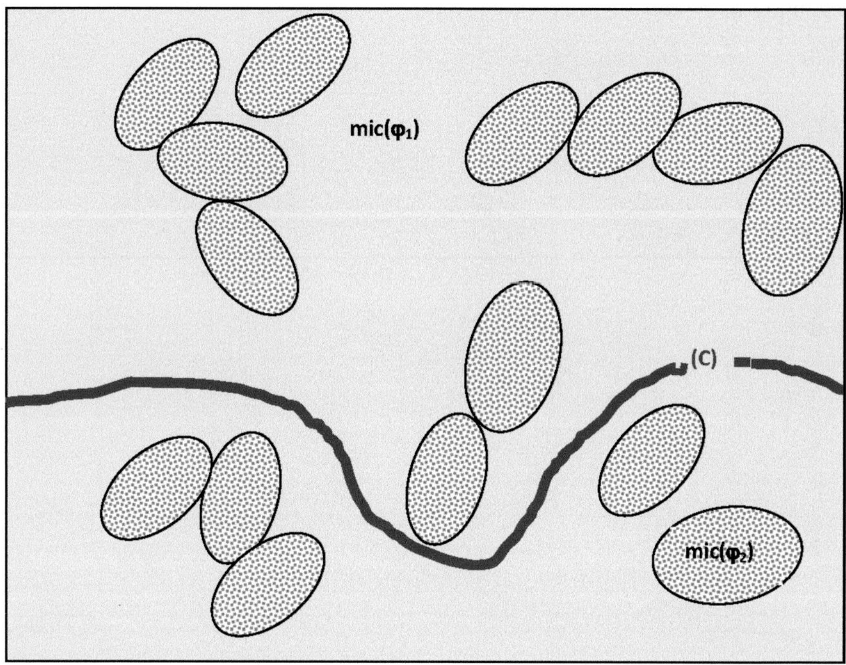

Fig. 2.2 Dilute Polyphasic System. In a dilute polyphasic system, there is a dispersing phase that features a path that makes it possible to travel through the entire phase without crossing an interface

indispensable to a certain independence regarding the spatial characteristics of the system and to the derivation (mathematically) of certain functions, etc. The option of choosing a level of description that makes it possible to treat the system as homogeneous, whereas in reality it is not (pseudohomogeneous) is the cornerstone of the definition of PDS. All mathematical theory and modeling that follow are based on this property of pseudohomogeneity. In reality, in the literature, the property of pseudohomogeneity is often implicit and considered as given. It follows that the relevance of representation or of modeling is not necessarily guaranteed.

2.2 Characteristic Volume and Pseudohomogeneity

The concept of pseudohomogeneity does not apply if a system can be divided into elements of volumes that are statistically equivalent from the point of view of their properties. These elements of volume must still be quite small in comparison with the total volume so that the medium, considered on a small scale, can appear as continuous. The same problem presents itself in physics when attempts are made to explain the macroscopic properties of a system (thermodynamics, for example) beginning with properties that are on a microscopic level (atoms, molecules …). In this case the dimensions of the particles are of the order of the angstrom (Å; 10^{-10} m) and their number is of the order of Avogadro's number (6.023×10^{23}). On this scale, a macroscopic system of a solution of HCl or of a gas in an equilibrium of let us say 1 cm^3 can be considered as perfectly homogeneous. It is the role of statistical physics to analyze the relationship between microscopic distributions and macroscopic properties, and this represents a considerable branch of physics. Obviously, the aim here is not to develop a sophisticated statistical theory, but simply to try to understand more quantitatively, the concept of pseudohomogeneity that is often accepted without saying.

If anyone has seen a suspension of baker's yeast, well stirred in a 250 ml conical flask, they will willingly concede that this beige, milky liquid of uniform color is homogenous. However, an elementary microscopic examination makes it appear as a heterogeneous system, formed of cell structures that are very different from the medium of the culture and are more or less regularly placed in the microscope's field. At this level of observation, there is no doubt that the system is heterogeneous while we are ready to consider some dozens of cm^3 of this suspension as homogenous.

The concept of homogeneity of a system could be defined in the following manner: two (or more) elements of equal volume sampled randomly from two (or more) points of the system, present exactly the same properties (physical, chemical, etc.). It is already known that this type of homogeneity does not exist in reality and the term "exactly" should be replaced by "statistically." So the relevant question is from what size two elements of volume can have statistically the same properties. To come back to the example of the suspension of yeasts, it seems inevitable that it should be considered, in reality, as heterogeneous but if a large volume is considered, it might be conceded that it can appear homogenous. On a certain scale, it can be said that it is pseudohomogenous. However, this property can be used to study the system only if the element of volume from which the suspension can be called pseudohomogenous is small in comparison with the whole system.

A "naive" step to estimate the size of this element of volume can be attempted. Take into consideration that each micelle of the system can be characterized by a value of between 0 and 1. In the case of a cell, it can be a question of, for example, its size characterized by the relationship of its present size and its maximum size before division; or the relationship with a quantity of constitutive enzymes at maximum level that it can present; or the relationship with the present number of several receptors in the membrane in relationship to the maximum quantity, etc., a characteristic size ξ (a property) is defined as:

$$\xi = \frac{x}{\max(x)}; \quad 0 \le \xi \le 1 \tag{2.2.1}$$

Let us take the equally probable case of each micelle having the same probability of finding itself in state ξ_1 as in state ξ_2. In other words, we consider that for ξ there is a uniform distribution. Over the interval [0, 1] the average value for ξ is $\bar{\xi} = 0.5$ and the variance is $\sigma_\xi^2 = 1/12$ (Kaufman 1965).

By numeric simulation, we have estimated the average and the variance of the property ξ in two elements of volume that have a growing number of micelles. Estimation of the average has been calculated by

$$\bar{\xi} = \frac{1}{n}\sum_{i=1}^{n}\xi_i$$

and that of the variance by,

$$s_{\xi}^{2} = \frac{1}{n}\left[\sum_{i=1}^{n}\xi_i^2 - \frac{1}{n}\left(\sum_{1}^{n}\xi_i\right)^2\right]$$

where n is the number of micelles contained in an element of volume (Dagnelie 1980, 1981).

Figures 2.3 and 2.4 show that the difference between average values for ξ obtained in the two volumes is significantly different up to 10^4 micelles per element of volume.

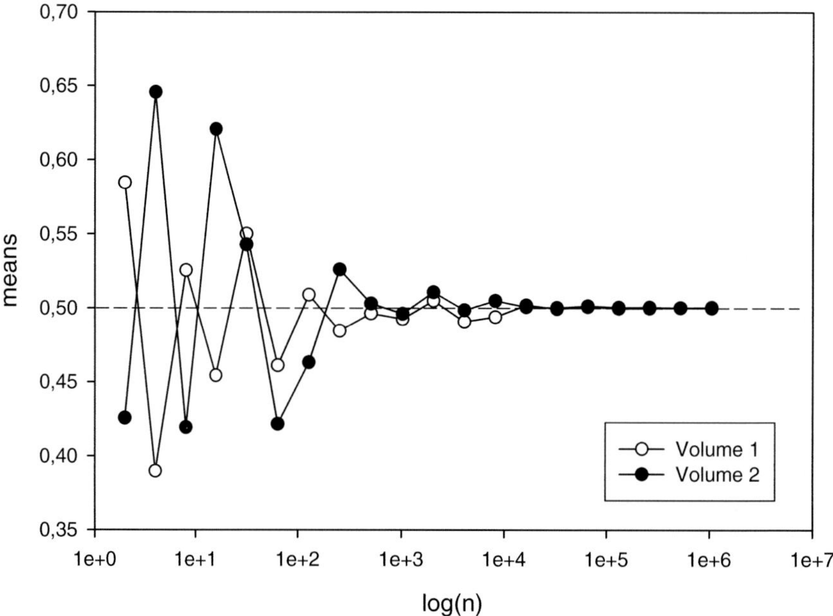

Fig. 2.3 Means according to the number of micelles. When average values of a micellian property distributed uniformly over the interval [0, 1] in two elements of volume (1 and 2) are calculated, very different average values are noted when the number of micelles (n) is less than or equal to 10,000. Around 10^5 micelles per element of volume, the two means converge toward the same value, corresponding to the theoretical value of the distribution (0.5). This value of 10^5 can be considered as the smallest number of micelles necessary for the average properties of the two elements of volume to be considered as statistically equal (The *dashed line* indicates the theoretical value of 0.5)

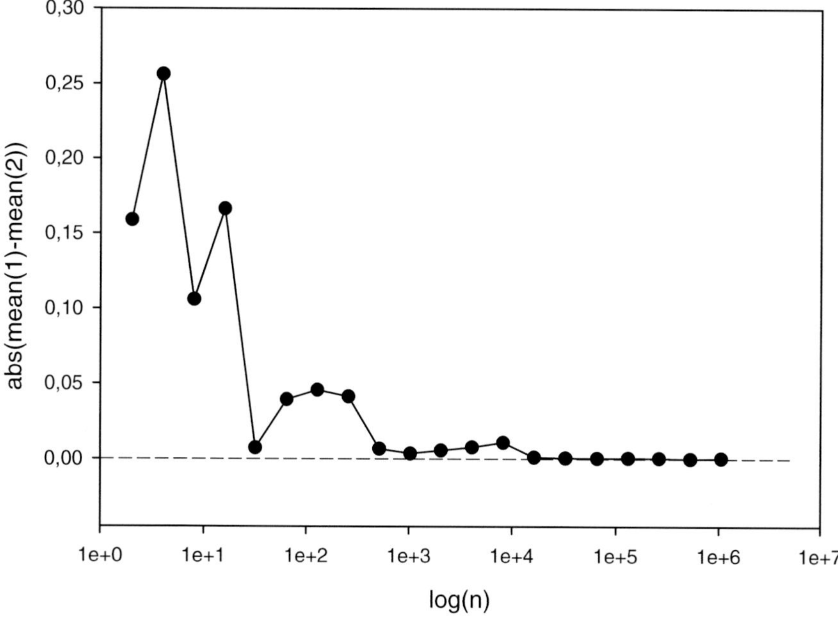

Fig. 2.4 Absolute values of the difference between the two means. The graph shows the absolute values for the difference between the two mean values of the preceding figure. The difference decreases rapidly with the number of micelles and can be considered as zero beyond 10^5 micelles by element of volume

Figure 2.5 shows that variance reaches its theoretical value in the two volumes toward about 10^5 particles. This is a value for which the average values are about the same and very near to 0.5.

So, in the case where each micelle has the same probability of having a property of between 0 and 1 (uniform distribution), there must be 10^5 micelles per element of volume for two of these elements of volume, chosen at random, to have the same property average. On this basis, the order of size can be sought of the smallest element of volume that satisfies the criteria of pseudohomogeneity. Obviously, the answer is not universal and depends on the system studied. Let us take, for example, a bioreactor of 1 l in which *Escherichia coli* is being cultivated. The mass of a cell is estimated to be 2.8×10^{-13} g (dry weight; Neidhardt et al. 1994). Let us imagine that the total biomass in the reactor is 1 g (dry weight). The system contains $M^c/m^c = 1/2.8 \times 10^{-13}$ some 3.6×10^{12} cells (obtained as the relationship between the total mass and the mass of one cell). One liter of culture contains this number of cells, and this makes it possible to calculate the volume $\Delta V*$ that contains 10^5 of them, the number necessary for pseudohomogeneity. Given that

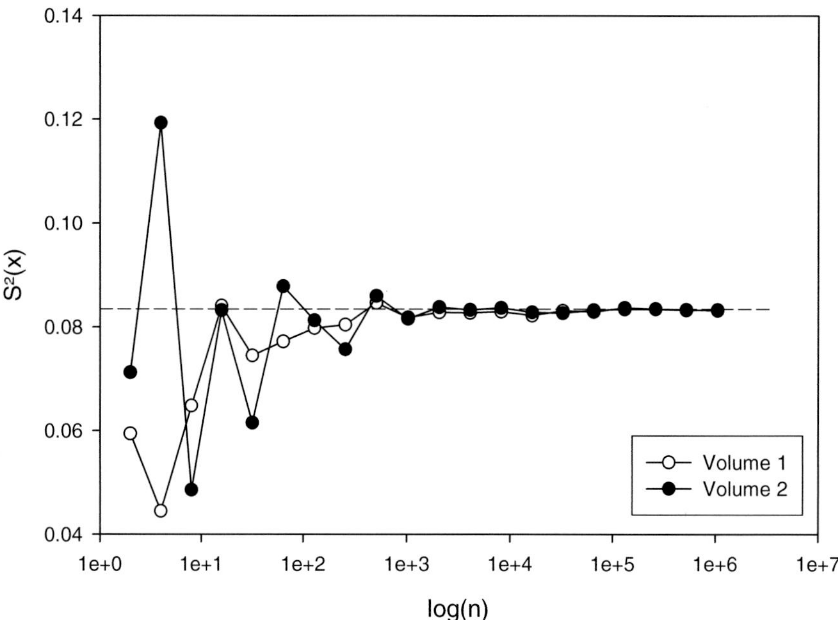

Fig. 2.5 Estimation of the variance according to the number of micelles. When the number of micelles is not raised to any extent, the variances of the property between the two elements of volume are different and reveal through this a greater heterogeneity in physiological states between micelles of one element of volume and another. The same applies in the case of the average (cf. Figs. 2.3 and 2.4), the variances converge toward the theoretical value (1/12; shown as *dashes*) when the number of micelles exceeds 10^5 by element of volume

$$\Delta V^* = \frac{V_T n^*}{N_T^c}$$

where V_T is the total volume, N_T^c, the total number of cells and n^* the number of cells necessary for pseudohomogeneity. It is found that $\Delta V^* = 10^5/3.6 \times 10^{12}$, if $\Delta V^* = 2.8 \times 10^{-8}$ L, which corresponds to about three hundred thousandths of a milliliter. The whole system is more than 35 million times bigger than the element of volume in consideration. Moreover, the cell volume is equal to

$$v^c = \frac{m_c}{\delta_c}$$

where δ_c is the density of a wet cell that is estimated at 1000 g/L (Dammel and Schroeder 1991; Kubitschek et al. 1983, 1984; Baldwin and Kubitschek 1984). The cell volume of a wet cell, with 70 % water (Neidhardt et al. 1994), is equal to

$2.8 \times 10^{-13}/(0.3 \times 1000)$ that comes to about 9×10^{-16} L. The element of volume is then 30 million times bigger than the cell. In this example, it can be incontestably confirmed that the element of volume is truly considerably greater than the micelle (cell) and sufficiently smaller than the whole system (the bioreactor) (By estimating the volume of *E. coli* to be 4×10^{-16} L, according to the date from Neidhardt et al. (1994), the element of volume is found to be 70 million times bigger).

If the thinking above shows that the concept of pseudohomogeneity is really applicable theoretically to suspend microorganism cultures, it is however necessary to introduce an important practical remark on the order of sizes calculated above. In fact, as has already been mentioned, the dispersed systems that are taken into consideration are unstable, in the sense that they only support themselves when dispersed if energy is brought in, in the form of agitation. The calculations above rest on the hypothesis (implied) that the system is perfectly mixed. There is not, in this case, any spatial distribution at all, any gradient, etc. It is on this one condition that the two elements of volume of the same size, taken at random can have identical average composition. Now, this condition of perfect mix, that is essentially a chemical problem, is extremely difficult to realize in practice. Numerous factors intervene (rotation speed, agitators shape, geometry of the reactor, viscosity of the medium, etc.) that makes the achievement of perfect mix very arduous. In reality, the perfect mix does not exist in a concrete system and is merely an ideal. It has been observed, in a suspension of activated sludge from a treatment plant that is subjected simultaneously to mechanical agitation (500 rpm; propeller rod stirrer) and to agitation by air bubbles (air flow of about 100 L/h through a 3 mm diameter tube, without diffuser), that there is an establishment of a vertical gradient of 0.5 gDW/l/m in a 10 l PTFE carboy with a column of water of the order of the meter in height. (The average value of biomass was 3 GDW/L.) The gradient represents a variation of about 17 % between the top and the bottom of the carboy. These remarks show just how difficult it is to master and how important the phenomena of agitation are. They also show that the element of volume calculated above is probably too small by several orders of size. Fortunately, the relationship $V_T/\Delta V^*$ is so high that this modification of ΔV^* remains without consequences in practice.

In conclusion, the very simplified results of the calculations confirm the hypothesis that a suspension of microorganisms can be considered conceptually as pseudohomogeneous in a device on laboratory or industrial scale. Nevertheless, when moving from the concept of an application, it is useful to make sure that all precautions are taken at the chemical engineering level to make sure that the system can be considered as a reasonably perfect mix. Unfortunately, in the literature, articles on laboratory scale give very little information on agitation in small bioreactors. When such examples will be treated, the hypothesis that the mixing is sufficient will be implied.

2.3 Reaction Concentration and Pseudohomogeneity

In a PDS, the micelle does not have to be homogeneous. They can be composite, structured, compartmentalized, etc. In particular, they can contain organized or soluble substances and be the site of physical and/or (bio)chemical changes, etc. The level of description on micelle level is called the local level of description. If the hypothesis that the laws of kinetics apply at this local level, is accepted, the relevant quantity of expression of these kinetics is an intensive quantity, namely the local concentration, defined by

$$C_i^p = \frac{m_i^p}{v^p} \tag{2.3.1}$$

where C_i^p is the local concentration of mass of the compound i in the micelle; m_i^p is the mass of the compound i in the micelle of phase p, and v^p is the volume of the micelle.

The local concentration is therefore a mass concentration (and not molar or any other) that is called reaction concentration or R-concentration. The R-concentrations are therefore the intensive values necessary to describe the kinetics at micellian level, but they are of little use at the system level. If the PDS is pseudohomogenous, the concept of partial pseudohomogenous concentration can be introduced and make sense (partial because it only concerns the product i in phase p) defined by

$$\tilde{C}_i^p = \frac{1}{V_T} \sum_{i=1}^{N_T^p} m_i^p \tag{2.3.2}$$

where $\tilde{C}_i^p$ is the partial pseudohomogenous concentration of i associated with phase p of the system; V_T the total useful volume of the system and N_T^p the total number of micelles in phase p, m_i^p keeping the same significance as before. From a chemical point of view, (2.3.2) expresses the concentration that would be measured if, by one procedure or another, all micelles in phase p had been destroyed to liberate compound i into the global volume of the system. This is really a virtual volume that extends throughout the system. For this reason, we call it E-concentration (extended concentration).

The concentration defined by (2.3.2) is called partial because it concerns only the mass of the compound i contained in phase p. This compound can, of course, be found in other phases. There is therefore a total pseudohomogenous concentration that is obtained by adding together all phases:

$$\tilde{C}_i = \frac{1}{V_T} \sum_{j=1}^{N_p} \sum_{i=1}^{N_T^p} m_i^j \tag{2.3.3}$$

where N_p is the total number of phases. It is clear that the sum of all phases and all micelles from m_i is nothing other than the mass of the compound i in the system that is defined by

$$M_i = \sum_{j=1}^{N_p} \sum_{i=1}^{N_T^p} m_i^j \quad (2.3.4)$$

The pseudohomogenous concentration is simply,

$$\tilde{C}_i = \frac{M_i}{V_T} \quad (2.3.5)$$

This latest relationship corresponds to the usual definition of mass concentration in a homogeneous system. If C_i^0 is this homogeneous concentration, it is evident that

$$C_i^0 = \tilde{C}_i \quad (2.3.6)$$

The homogenous mass concentration and the total pseudohomogenous concentration are therefore represented by the same numeric value, but the physicochemical signification of the two quantities differs greatly. Moreover, by combining (2.3.2) and (2.3.3), it is easily shown that

$$\tilde{C}_i = \sum_{j=1}^{N_p} \tilde{C}_i^j \quad (2.3.7)$$

That simply expresses the total pseudohomogenous concentration is the sum of the partial pseudohomogenous concentrations (sum over all the phases).

2.3.1 Consequences of the Definition of Pseudohomogeneity

The definition of pseudohomogeneity that we have used has important consequences as regards the manner in which various quantities are calculated. We have introduced the concept of critical element of volume in such a way that two elements of volume are statistically identical. This choice allows the definition of significant average values from the center of an element of volume and the elimination of the necessity of using a structured representation of micelles. Let us take a simple example. The time that passes between the moment of cell division and the moment in mind is called "cell age." It is evident that in each element of volume, there are cells of different ages, the ages being distributed according to a certain statistical law. The models that take into account this type of distribution are called structured (refer, for example, to Fredrickson and Tsuchiya 1963,

Tsuchiya et al. 1966, Minkevitch and Abramychev 1994, Schügerl and Bellgardt 2000). In our approach, the elements of volume are chosen so that the average age of the cells is the same in all elements of volume and the same goes for all micellian properties. In fact, it is considered that in an element of volume of convenient size, all cells are identical and characterized by average values. Let us examine the impact of this point of view on the following relationship

$$M_i^p = \sum_{i=1}^{N_T^p} m_i^p \tag{2.3.8}$$

that expresses that the whole mass of the compound i in phase p is equal to the sum of the masses of this compound contained in all micelles of the phase. This relationship is evidently always true, whatever the distribution of the compound in the various micelles. If it is now considered that all micelles are statistically identical m_i^p is then the average value of the compound in an average micelle. It can then be written that if N_T^p is the total number of micelles in the phase p, that

$$M_i^p = N_T^p m_i^p. \tag{2.3.9}$$

Just the opposite holds in (2.3.8), this relationship is only true in certain conditions that are exactly those that have been defined for the critical element of volume and the criteria for pseudohomogeneity.

This approach can be applied to various quantities, thus the volume of phase p will be

$$V^p = N_T^p v^p \tag{2.3.10}$$

That is to say N_T^p times the micellian volume v_p.

2.3.2 Phase Density

By definition phase density shall be called the quantity,

$$X^p = \frac{M^p}{V_T} = \frac{N_T^p m^p}{V_T} \tag{2.3.11}$$

That is the relationship of the total mass of phase p to the total useful volume.

According to custom, the term, "biomass" will be kept for this quantity when the micelles that are in mind in (2.3.11) are cells.

2.3.3 Phase Volume Relationship

Equation (2.3.11) can be rewritten as

$$N_T^p = \frac{X^p V_T}{m^p} \tag{2.3.12}$$

and (2.3.10) put in the form

$$N_T^p = \frac{V^p}{v^p} \tag{2.3.13}$$

By equalizing these two equations and rearranging, the following is found,

$$\frac{V^p}{V_T} = \frac{X^p}{\delta_p} \tag{2.3.14}$$

or, by definition,

$$\delta_p = \frac{m^p}{v^p} = \frac{M^p}{V^p} \tag{2.3.15}$$

is the mass volume of an average micelle or of a phase.

2.3.4 Relationship Between R- and E-Concentrations

Relationships above are going to make it possible to calculate the relationship between the R- and the E-concentrations.

By definition, the R-concentration is (cf. (2.3.1))

$$C_i^p = \frac{m_i^p}{v^p} = \frac{M_i^p}{V^p} \tag{2.3.16}$$

By multiplying the top right and bottom right of the equation by V_T and using (2.3.2), (2.3.8) and (2.3.14), it is easily shown that

$$C_i^p = \tilde{C}_i^p \frac{\delta_p}{X^p} \tag{2.3.17}$$

This equality shows the relationship that exists between the local reaction level (the R-concentrations) and the extended concentrations (E-concentrations). It is not directly implied that the phase density intervenes as a factor. This relationship (and its consequences) will be abundantly used in the application of explicit forms of mass balances of PDS (cf. Chap. 3, Sect. 3.2.3).

2.4 Mass Balances

Let us consider a micelle from phase p encircled by three phases $\varphi_j, \varphi_k, \varphi_l$, as represented in Fig. 2.6.

The micelle that belongs to phase p is an open system (exchanging energy and matter with each of the phases). Let us assume that each phase contains a constituent *Pi*, represented by its index i. The transfer flux of the compound i (expressed as a unit of mass of the compound per unit of time, for example, in g/h) is represented by $\phi^x_{i,q}(p)$, where i is the index for the transferred compound, q, the index for the phase of origin and p that of the phase concerned (that of the micelle). The exponent x can take value E or S and these indicate whether the compound is entering or leaving the micelle. Moreover, the compound can undergo changes (biochemical, physical, etc.) inside the micelle. These changes are represented by the reaction rate per unit of volume, $r^p_i(.)$ where i is the reacting compound index, p the phase where the change is taking place; (.) indicates that the kinetic r is a complex function of several variables (concentrations, kinetic coefficients, temperature, pH,…). In a general way, $r^p_i(.)$ is a resulting kinetics which can be broken up into several terms (possibly with different signs):

$$r^p_i(.) = \sum_m r^p_{i,m}(.) \tag{2.4.1}$$

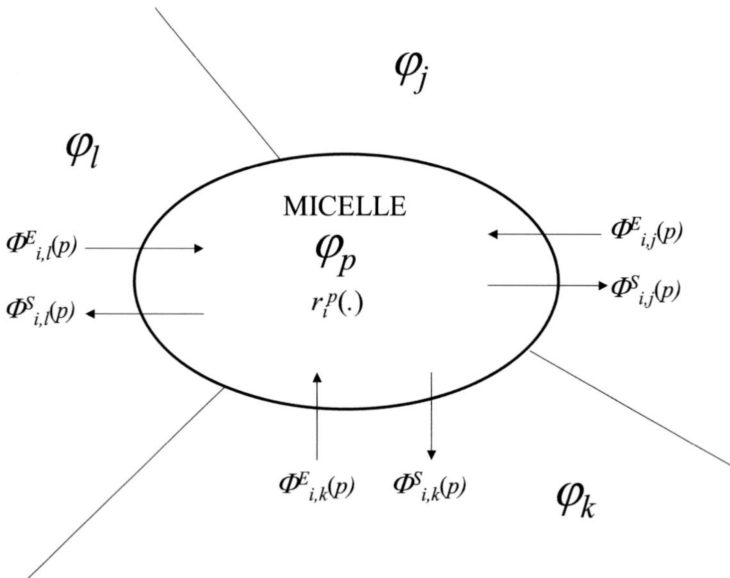

Fig. 2.6 Exchange and transformation terms. The micelle in phase p can exchange product i with the three surrounding phases j, k, and l. Transformation reactions $r^p_i(.)$ take place within the micelle

On the micelle level, kinetics depend (among others) on local concentrations, that is to say, the R-concentrations. Formally,

$$r_i^p(.) = (\{C_n^p\}, \ldots) \tag{2.4.2}$$

In accordance with usual kinetics, the units of r are expressed in mass units per unit of time and per unit of micellian volume (for example, in g/(h L) or in g/(h μm^3)).

The mass balance of the compound i at the center of the micelle is written as

$$\frac{\mathrm{d}m_i^p}{\mathrm{d}t} = \sum_{q=1;q\neq p}^{N_p} \left[\phi_{i,q}^E(p) - \phi_{i,q}^S(p)\right] + r_i^p(.)v^p \tag{2.4.3}$$

where m_i^p is the mass of the compound i in the micelle and v^p is the average volume of the micelle. The first term of the right side represents the total transfer flows in all phases (N_p is the number of phases).

In order to simplify the written formula, it is written as,

$$\varphi_{i,\{q\neq p\}}(p) = \sum_{q\neq p} \left[\phi_{i,q}^E(p) - \phi_{i,q}^S(p)\right] \tag{2.4.4}$$

This quantity represents the net interphasic exchange flux of the compound i at the level of the micelle. The balance (2.4.3) is written as,

$$\frac{\mathrm{d}m_i^p}{\mathrm{d}t} = \phi_{i,\{q\neq p\}}(p) + r_i^p(.)v^p \tag{2.4.5}$$

From the point of view taken, the balance at micelle level is not useful; a more macroscopic description will be sought. Let us begin by writing the mass balance for the total number of micelles that make up phase p.

Let N_T^p be the number of micelles that make up the phase. Since all micelles are statistically identical, simply multiply the left and right side of Eq. (2.4.5) by N_T^p; therefore,

$$N_T^p\frac{\mathrm{d}m_i^p}{\mathrm{d}t} = N_T^p\phi_{i,\{q\neq p\}}(p) + N_T^p r_i^p(.)v^p \tag{2.4.6}$$

$N_T^p v^p$ is nothing other than the volume of the phase,

$$V^p = N_T^p v^p \tag{2.4.7}$$

Moreover,

$$\Phi_{i,\{q\neq p\}}(p) = N_T^p \phi_{i,\{q\neq p\}}(p) \tag{2.4.8}$$

The total mass of the compound i for all the micelles of this phase is simply

$$M_i^p = N_T^p m_i^p \tag{2.4.9}$$

In a general case, the number of micelles is not constant and the derivative with respect to time of (2.4.9) is,

$$\frac{\mathrm{d}M_i^p}{\mathrm{d}t} = N_T^p \frac{\mathrm{d}m_i^p}{\mathrm{d}t} + m_i^p \frac{\mathrm{d}N_T^p}{\mathrm{d}t} \tag{2.4.10}$$

By using the relationships, (2.4.7) to (2.4.10) in (2.4.6), the mass balance of the product i over the group of micelles is written as,

$$\frac{\mathrm{d}M_i^p}{\mathrm{d}t} = \Phi_{i,\{q\neq p\}}(p) + r_i^p(.)V^p + m_i^p \frac{\mathrm{d}N_T^p}{\mathrm{d}t} \tag{2.4.11}$$

By dividing and multiplying the last term of the right side by N_T^p and by using the logarithmic derivative, it is finally found that

$$\frac{\mathrm{d}M_i^p}{\mathrm{d}t} = \Phi_{i,\{q\neq p\}}(p) + r_i^p(.)V^p + M_i^p \frac{\mathrm{d}\ln N_T^p}{\mathrm{d}t} \tag{2.4.12}$$

It could be believed that the balance (2.4.12) truly is the required expression of the phase balance. However, it should be mentioned that nothing has been said about the real apparatus in which the phases and micelles are contained. In reality, the balance (2.4.12) is complete for a closed apparatus; it is incomplete for an open apparatus. Since the main interest is in an open system (as the chemostat), it is convenient to complete the balance. The incompleteness of (2.4.12) is not, however, immediately taken notice of nor implied, because the variational term for the number of micelles ($\mathrm{d}_t \ln N_{Tp}$) leaves it to be assumed that the system is open. It is no such thing, in fact, because this term appears only as the variation in micelles containing the compound i. In reality, the balance is closed in relation to the compound, in spite of the variational term of micelles. In fact, the term $\mathrm{d}_t \ln N_{Tp}$ does not represent an inflow/outflow term for micelles, but only expresses the combining aspect of the interphasic exchange process of i and the way the compound is distributed between the different micelles.

To complete the balance, the level of description chosen should be above that for the micelle and should include the entire system.

At system level, the variation in mass of the compound can be written as,

$$\frac{\mathrm{d}M_i}{\mathrm{d}t} = F_i^E - F_i^S + \sum_p \text{contribution of the micella} \tag{2.4.13}$$

where F_i^E and F_i^S are global terms for inflow of i via one or several phases (including possibly the dispersing phase). The sum of the contribution of the micelles is calculated for all phases (it being understood that phases that do not contain i have no contribution).

By using (2.4.12), it is found that

$$\frac{\mathrm{d}M_i}{\mathrm{d}t} = F_i^E - F_i^S + \sum_p \left[\Phi_{i,\{q\neq p\}}(p) + r_i^p(.)V^p + M_i^p \frac{\mathrm{d}\ln N_T^p}{\mathrm{d}t}\right] \tag{2.4.14}$$

The term $\Phi_{i,\{p\neq q\}}(p)$ represents the net transfer flow of the contributions from all phases of phase p. It is evident that the same quantity of i that enters into phase p leaves phases $q \neq p$. It is obvious that the reasoning applied to phase p can also be applied to all other phases. Ultimately, it is easy to understand that, at the level of the whole system, the interphasic transfer fluxes are canceled out, since all quantities of matter that exit a phase are necessarily compensated for by the inflow of matter in other phases and vice versa. So,

$$\sum_{p=1}^{N_p} \Phi_{i,\{q\neq p\}}(p) \equiv 0 \tag{2.4.15}$$

(A stricter equation is given as evidence in Appendix A.4.) In equation form, (2.4.14) is reduced to

$$\frac{\mathrm{d}M_i}{\mathrm{d}t} = F_i^E - F_i^S + \sum_p \left[r_i^p(.)V^p + M_i^p \frac{\mathrm{d}\ln N_T^p}{\mathrm{d}t}\right] \tag{2.4.16}$$

This relationship expresses the variation in mass of the compound i within the whole system. It includes, thanks to the inflow/outflow terms, the possibility of modifying arbitrarily, the number of micelles in each phase.

Before seeking the complete balance at the level of phase, this mass balance (in extensive form) will be expressed in an intensive form, that is to say, in terms of concentrations. For this, let us divide the left and right sides of (2.4.16) by V_T:

$$\frac{1}{V_T}\frac{\mathrm{d}M_i}{\mathrm{d}t} = \frac{F_i^E - F_i^S}{V_T} + \sum_p \left[r_i^p(.)\frac{V^p}{V_T} + \frac{M_i^p}{V_T}\frac{\mathrm{d}\ln N_T^p}{\mathrm{d}t}\right] \tag{2.4.17}$$

In a general case V_T is not a constant and

$$\frac{\mathrm{d}}{\mathrm{d}t}\left(\frac{M_i}{V_T}\right) = \frac{1}{V_T}\frac{\mathrm{d}M_i}{\mathrm{d}t} - \frac{M_i}{(V_T)^2}\frac{\mathrm{d}V_T}{\mathrm{d}t} \tag{2.4.18}$$

So,

$$\frac{\mathrm{d}}{\mathrm{d}t}\left(\frac{M_i}{V_T}\right) = \frac{F_i^E - F_i^S}{V_T} + \sum_p \left[r_i^p(.)\frac{V^p}{V_T} + \frac{M_i^p}{V_T}\frac{\mathrm{d}\ln N_T^p}{\mathrm{d}t}\right] - \frac{M_i}{V_T}\frac{\mathrm{d}\ln V_T}{\mathrm{d}t} \tag{2.4.19}$$

By using the definitions (2.3.1) and (2.3.5) for concentrations (2.4.19) it can be written as

$$\frac{\mathrm{d}\tilde{C}_i}{\mathrm{d}t} = \frac{F_i^E - F_i^S}{V_T} + \sum_p \left[r_i^p(.)\frac{V^p}{V_T} + \tilde{C}_i^p\frac{\mathrm{d}\ln N_T^p}{\mathrm{d}t}\right] - \tilde{C}_i\frac{\mathrm{d}\ln V_T}{\mathrm{d}t} \tag{2.4.20}$$

Taking into account that the total pseudohomogenous concentration is the sum of the partial pseudohomogeneous concentrations (2.3.7), (2.4.20) is expressed in terms of partial concentrations,

$$\sum_p \frac{\mathrm{d}\tilde{C}_i^p}{\mathrm{d}t} = \frac{F_i^E - F_i^S}{V_T} + \sum_p \left[r_i^p(.)\frac{V^p}{V_T} + \tilde{C}_i^p\frac{\mathrm{d}\ln N_T^p}{\mathrm{d}t}\right] - \sum_p \tilde{C}_i^p\frac{\mathrm{d}\ln V_T}{\mathrm{d}t} \tag{2.4.21}$$

So it is easy to write the equation for each of the phases p by distributing properly the terms of (2.4.21):

$$\frac{\mathrm{d}\tilde{C}_i^p}{\mathrm{d}t} = \frac{\tilde{F}_i^{p,E} - \tilde{F}_i^{p,S}}{V_T} + r_i^p(.)\frac{V^p}{V_T} + \tilde{C}_i^p\left(\frac{\mathrm{d}\ln N_T^p}{\mathrm{d}t} - \frac{\mathrm{d}\ln V_T}{\mathrm{d}t}\right) + FE_i^p \tag{2.4.22}$$

The first term of the right side of (2.4.22) expresses the distribution of the inflows of each phase. These flows are expressed in terms of pseudohomogenous concentration from which the formulae are derived (tilde). The latest term of the right side expresses the interphasic exchange flows that appear when the system is divided into its distinctive phases.

FE_i^p is nothing other than the interphasic exchange term divided by the useful volume, namely,

$$FE_i^p \equiv \frac{\Phi_{i,\{p\neq q\}}(p)}{V_T} \tag{2.4.23}$$

and

$$\Phi^0_{i,\{p\neq q\}}(p) \equiv \frac{\Phi_{i,\{p\neq q\}}(p)}{V_T} \tag{2.4.24}$$

That is called the interphasic exchange flux by unit of volume.

By using the volumic relationship of phase (2.3.14) and writing,

$$q_i^p = \frac{r_i^p(.)V^p}{M^p} \tag{2.4.25}$$

The mass balance of the phase (2.4.22) is obtained in the following intensive form,

$$\frac{\mathrm{d}\tilde{C}_i^p}{\mathrm{d}t} = \frac{\tilde{F}_i^{p,E} - \tilde{F}_i^{p,S}}{V_T} + q_i^p X^p + \Phi^0_{i,\{q\neq p\}}(p) + \tilde{C}_i^p\left(\frac{\mathrm{d}\ln N_T^p}{\mathrm{d}t} - \frac{\mathrm{d}\ln V_T}{\mathrm{d}t}\right) \tag{2.4.26}$$

The left side expresses the time variation in pseudohomogenous concentration (E-concentration) of the compound i in phase p (unit: mass/(volume.time)). On the right side of (2.4.26):

- The first term expresses, in E-concentrations, the inflow and outflow terms of the compound i contained in phase p at the level of the global system. These are therefore inflow/outflow terms associated with the micelles of phases p (units: mass/(volume.time));
- The second term is the product of density of phase by the net kinetics of the changes in the compound at the center of the micelles of phase p. The phase density is essentially a mass concentration (mass/volume). Since q_i^p being speed per unit of phase density, it is in reality a specific speed (units: time^{-1});
- The third term is the net interphasic exchange flow (units: mass/(vol temps). It represents the net transfer flux of compound i stemming from or leaving all other phases;
- The fourth term is a variational term that takes into account the combined variation of the number of micelles in the phase and the variation in the useful volume. For a system of constant volume, this latest contribution is obviously nil.

Note: It appears clearly in this section that the decomposition of a global system into its different phases (Eq. 2.4.22) makes the interphasic exchange flows appear and these are not reduced to equations in a description based only at the level of the system.

2.4.1 Note Concerning the Variational Term Appearing in (2.4.12)

In order to keep as general a character as possible, it is assumed that the number of micelles was not constant at the time of the derivation of (2.4.9). This introduces the number of micelles time derivative and is called the variational term. It is important to understand the import of this term to avoid misinterpretation in the future, mainly at the time of derivation of the rate of everything (refer to Appendix A.3). To remind us, in the process of derivation of (2.4.12), it was begun with a mass balance at the level of the micelle to construct, by addition, the balance at the level of the phase. Variation in the number of micelles included in the variational term is therefore a variation at the phase level only and does not represent an inflow/outflow term at the global level of the whole system. Even more important, the variational term does not represent the variation in the number of micelles by division that cell division alone produces in general, nor by cell division in particular. In fact, it is easy to understand that cell division neither produces nor consumes any compounds at all (except a minimum part linked to the energy necessary for the division). A representation of cell division alone cannot be brought in as a source or sink term for a balance such as (2.4.10). In short, it is known that the variational term does not exist for it is neither a term of exchange with the exterior world, nor a term representing an aggregation or the division of micelles (p. ex. cell division). In reality, this term represents an interphasic exchange term, that is to say, the inflow or the outflow of micelles in phase p and stemming from one (or several) other phases. Let us consider a concrete example. Let us imagine a solid phase composed of two cell compartments, namely inactive cells A and active cells A^*. If an activation/inactivation process of the type $A \leftrightarrows A^*$ exists, account must be taken of this variation in the number of micelles in phase p and in phase p^*. It is this type of situation that takes into account the variation term that appears in (2.4.10) and (2.4.12). If there is only one single cell phase, a cell viability rate of 100 % implies that the variational term is zero; if there is cell mortality, the variational term is necessarily negative, implying a loss of micelles from phase p by cell death. In this latter case, there is no interphasic exchange of micelles to speak of since there is only one micellian phase, but there is nevertheless an exchange flow for the dispersing phase. (If this state of affairs is annoying as regards formulae, it is always possible to create a compartment of dead cells and so generate a second micellian phase.)

A slightly different way of approaching this delicate problem is described in Appendix A.5.

2.5 The Grouping Principle

Equation (2.4.26) is valid for a specific compound. However, it is sometimes desirable to consider groups of compounds (or families) of particular interest.

Let NC^p be the total number of different constituents of phase p and let NF^p be the number of families of compounds from this same phase ($NF^p \leq NC^p$). Now let NC_k^p be the number of different compounds belonging to the family (or group) k from the phase p ($NC^p = \sum_k NC_k^p$).

Since the E-concentrations are additional, a partial, pseudohomogenous concentration can be defined by,

$$\tilde{C}_k^p = \sum_{i=1}^{NC_k^p} \tilde{C}_i^p; \quad i \in \Im_k \tag{2.5.1}$$

where $\Im_k$ represents the family or the group k. So,

$$\frac{\mathrm{d}\tilde{C}_k^p}{\mathrm{d}t} = \sum_{i=1}^{NC_k^p} \frac{\mathrm{d}\tilde{C}_i^p}{\mathrm{d}t}; \quad i \in \Im_k \tag{2.5.2}$$

and the laws of evolution, themselves, are also additive.

Consequently, the system of equations NC^p in each phase p can be replaced by a reduced system of equations NF^p in the form,

$$\frac{\mathrm{d}\tilde{C}_k^p}{\mathrm{d}t} = \frac{1}{V_T} \sum_{i=1}^{NC_k^p} (\tilde{F}_i^E - \tilde{F}_i^S) + X^p \sum_{i=1}^{NC_k^p} q_i^p + \sum_{i=1}^{NC_k^p} \Phi_{i,\{q \neq p\}}^0(p) + \tilde{C}_k^p \left(\frac{\mathrm{d}\ln N_T^p}{\mathrm{d}t} - \frac{\mathrm{d}\ln V_T}{\mathrm{d}t} \right) \tag{2.5.3}$$

which is written in an identical form in (2.4.26)

$$\frac{\mathrm{d}\tilde{C}_k^p}{\mathrm{d}t} = \frac{1}{V_T} (\tilde{F}_k^E - \tilde{F}_k^S) + q_k^p X^p + \Phi_{k,\{q \neq p\}}^0(p) + \tilde{C}_k^p \left(\frac{\mathrm{d}\ln N_T^p}{\mathrm{d}t} - \frac{\mathrm{d}\ln V_T}{\mathrm{d}t} \right) \tag{2.5.4}$$

with

$$q_k^p = \sum_i q_i^p; \quad \tilde{F}_k^E - \tilde{F}_k^S = \sum_i (\tilde{F}_i^E - \tilde{F}_i^S); \quad \Phi_{k,\{q \neq p\}}^0(p) = \sum_i \Phi_{i,\{q \neq p\}}^0(p) \tag{2.5.5}$$

In other terms, the evolution law for a family (or a group) of compounds is strictly isomorphic to the evolution law of an isolated compound. This result is particularly interesting in the sense that it makes it possible to introduce several levels of description for the center of the system, using the same formulae. The evolution of the system can then be followed at the metabolite level or just that the level of groups of metabolites (by family, group, or metabolic pathway) or by combinations of each of these levels of description.

2.6 Comment on a Certain Relativity in Writing the Mass Balance

It can be quite annoying to note that it is necessary to use two levels of description (micellian and systemic) to obtain the correct and complete expression of the balance in a phase.

On one hand, if one starts at the micellian level of description and if we sum on them, the phase balance is given by (2.4.12). The balance of the system is then given by the sum of the phase balances over all the phases.

On the other hand, by starting at the system level of description, the phase balance is obtained by decomposition of the whole system into its various phases.

What is remarkable is that it is by using these two methods that are apparently symmetrical, the results obtained are not the same.

Let us quickly go over the development again by distinguishing the levels of description used. The relationship (2.4.14) can be put in the form,

$$\left.\frac{\mathrm{d}M_i}{\mathrm{d}t}\right|_{\mathrm{SYS}} = F_i^E - F_i^S + \left.\frac{\mathrm{d}M_i}{\mathrm{d}t}\right|_{\mathrm{mic}} \tag{2.6.1}$$

The notation $|_X$ indicates that the level of description used to derive the expression ("level X" in the example).

By dividing the left and right sides of (2.6.1) by a constant volume V_T, it is obtained that

$$\left.\frac{\mathrm{d}\tilde{C}_i}{\mathrm{d}t}\right|_{\mathrm{SYS}} = \frac{F_i^E - F_i^S}{V_T} + \left.\frac{\mathrm{d}\tilde{C}_i}{\mathrm{d}t}\right|_{\mathrm{mic}} \tag{2.6.2}$$

The phase balance obtained from micelles is given by (2.4.12); all that needs to be done is to add up all the phases to obtain the contribution at the level of the system. So,

$$\left.\frac{\mathrm{d}\tilde{C}_i}{\mathrm{d}t}\right|_{\mathrm{mic}} = \sum_p \left.\frac{\mathrm{d}\tilde{C}_i^p}{\mathrm{d}t}\right|_{\mathrm{mic}} \tag{2.6.3}$$

On one hand, (cf. (2.3.7))

$$\tilde{C}_i = \sum_p \tilde{C}_i^p \tag{2.6.4}$$

and on the other hand, the global inputs/outputs of i can be distributed between the different phases,

$$F_i^E - F_i^S = \sum_p \left(\tilde{F}_i^{p,E} - \tilde{F}_i^{p,S} \right) \quad (2.6.5)$$

Using (2.6.5) in (2.6.2), it is obtained that

$$\sum_p \left. \frac{d\tilde{C}_i^p}{dt} \right|_{SYS} = \sum_p \frac{F_i^{p,E} - F_i^{p,S}}{V_T} + \sum_p \left. \frac{d\tilde{C}_i^p}{dt} \right|_{mic} \quad (2.6.6)$$

It is easy to distribute the terms of the sum by letting each affect a phase and so,

$$\left. \frac{d\tilde{C}_i^p}{dt} \right|_{SYS} = \frac{F_i^{p,E} - F_i^{p,S}}{V_T} + \left. \frac{d\tilde{C}_i^p}{d} t \right|_{mic} \quad (2.6.7)$$

where the latest term from the right side is no other than (2.4.12) divided by V_T. It is shown that

$$\left. \frac{d\tilde{C}_i^p}{dt} \right|_{mic} = \Phi^0_{i,\{p \neq q\}}(p) + r_i^p(.) \frac{V^p}{V_T} + \tilde{C}_i^p \frac{d \ln N_T^p}{dt} \quad (2.6.8)$$

Let us write, (2.6.7) in the more compact form,

$$v(i)|_{SYS} = v_{E/S}(i) + v(i)|_{mic} \quad (2.6.9)$$

where $v(i)|_{SYS}$ and $v(i)|_{mic}$ are rates of changes in i obtained, respectively, by a level of description at the level of the system and the micelle, $v_{E/S}(i)$ is the net inflow/outflow speed of the system. By analogy, the term "referential" can be used rather than level of description. It is then noted that the speed of (bio)chemical change of the compound i in phase p depends on the referential used when writing the equations. This proposition is to be matched with this: "*In different basic systems the laws governing movement do not generally take the same form.*" (Landau and Lifchitz 1969), that is a basic proposition in the Galilean principle of relativity.

The important point about this principle of kinetic relativity (PKR) is that the relativity of the observed rates depends on whether or not the system is open. In fact, if the system is closed, $v_{E/S}(i) \equiv 0$ and so from (2.6.9):

$$v(i)|_{SYS} = v(i)|_{mic} \quad (2.6.10)$$

So the choice of referential makes no difference.

2.6.1 Discussion

It is not thought that this principle of kinetic relativity or PKR is of just small import from the point of view of current practice and of mainly the modeling of cell mechanics and/or bioreactors. In the first case, it happens frequently that cell physiological models are developed without specifying which environment the cell is in (batch, fed-batch, chemostat, etc.). It is possible to write a model in these conditions, but it is not possible to complete experimental verification without choosing a particular environment for the cell. The PKR can then cause discordances between the model and experimental verification, to appear simply because the ways in which the model is obtained belong to one referential, whereas the experimental verification belongs to another. Just the same, the observation of state variables at the level of a bioreactor can lead to false deductions about intracell mechanisms if possible corrections due to PKR are ignored. Figure 2.7 schematizes the situation that leads to the relative speeds being obtained (2.6.9).

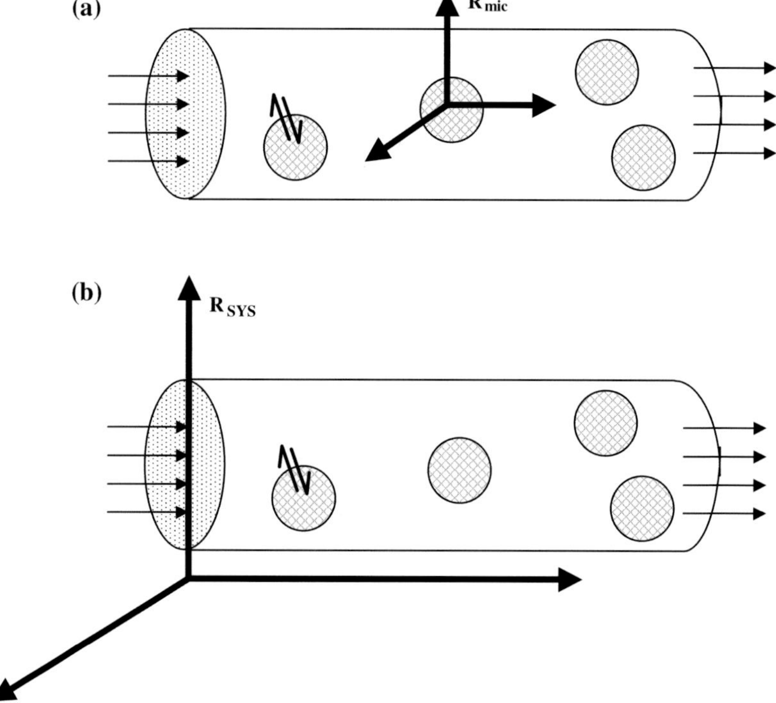

Fig. 2.7 Representation of the referentials. The *horizontal cylinder* represents an open system in which a flow causes micelles containing the compound i to get in and out. In case **a**, the kinetics of the transformation in i is obtained by fixing the referential on one micelle (R_{mic}). In case **b**, the kinetic is obtained by working within the referential of the whole system (R_{SYS}). The results obtained are different and linked by the relationship (2.6.9). By analogy with the mechanics, in R_{mic} the micelles is "at rest" (corresponds to a closed system), whereas in R_{SYS}, the micelle is "in motion" (corresponds to an open phase)

2.6.2 Example

To reinforce the considerations above of the possibility of a PKR, a very simple example of relativity based on a more intuitive than mathematical approach is proposed.

Figure 2.8 shows the development of biomass X (cell phase density) during the sudden increase in the dilution rate in a chemostat previously at steady state.

For t < 0, a steady state has been reached where $D = D_1$; at t = 0, there is a sudden shift up caused by rapidly increasing D to D_2, with $D_2 > D_1$. A transient regime follows that is characterized by a decrease in biomass, and this regime tends toward another stationary state (if D_2 does not exceed the washout value).

Let us assume that there is an analytical expression, X(t). In fact, in this example, a plausible arbitrary expression has been used (inspired by data described in Thierie et al. 1999):

$$X(t) = X(0)\left[\mathrm{e}^{-D_2 t} + b(1 - \mathrm{e}^{-kt})\right] \tag{2.6.11}$$

where $b = 0.1$ and $k = 0.5$; $D_2 = 0.6$ et $D_1 = b\ k = 0.05$ (2.6.11) is a positive, decreasing monotone function.

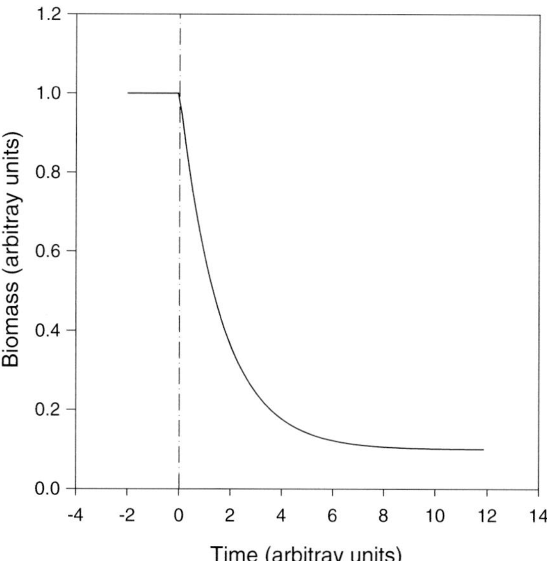

Fig. 2.8 Development in biomass after a dilution rate shift up. A steady state has been reached in the chemostat (time < 0) for a given dilution rate (D_1) when there is a sudden increase at time $t = 0$. The biomass decreased rapidly and relaxed to another steady state, corresponding to dilution rate D_2

At the start of such data, the only way to calculate the specific growth rate, μ is to calculate the ratio between the rate (time derivative) and the biomass, let

$$\mu = \frac{X^{/}(t)}{X(t)} \tag{2.6.12}$$

A negative specific speed of growth is obtained, since (2.6.11) is a monotonously decreasing function.

In reality, (2.6.12) is only the expression of net rate obtained at the system level and it should be written as

$$\mu|_{\mathrm{SYS}} = \frac{X^{/}(t)}{X(t)} \tag{2.6.13}$$

which is represented as dashes in Fig. 2.9.

The relationship (2.6.13) represents, in fact the balance between cell growth and the outflow of cells from the chemostat. So it is seen that the terms of inflow/outflow appear and thus intervene in the principle of kinetic relativity.

The passive outflow term can easily be evaluated, that is to say the outflow speed of cells in the absence of their growth. In a perfectly agitated system, the outflow speed will be equal to the dilution rate multiplied by the biomass in the reactor, let

$$\frac{\mathrm{d}X}{\mathrm{d}t} = -D_2 X \tag{2.6.14}$$

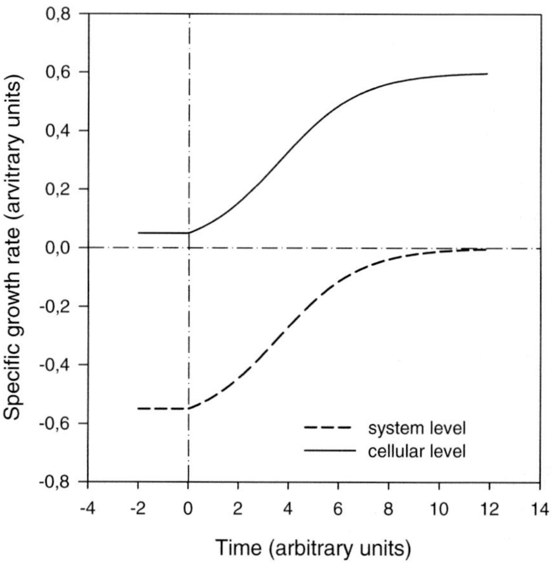

Fig. 2.9 Relative rates. The specific rate observed in the system referential (*in dashes*) is negative and tends toward zero, whereas the specific growth rate of the cells (micellian referential) is positive and tends toward its steady value, D_2

By rearranging, the specific speed is obtained,

$$\frac{1}{X}\frac{\mathrm{d}X}{\mathrm{d}t} = -D_2 \quad (2.6.15)$$

so, by definition,

$$\mu_{E/S} = -D_2 \quad (2.6.16)$$

By using (2.6.9), it is then possible to calculate the specific speed of growth of the cells (in the micellian referential),

$$\mu|_{\mathrm{cell}} = \mu|_{\mathrm{SYS}} - \mu_{E/S} \quad (2.6.17)$$

So by using (2.6.13) and (2.6.16)

$$\mu|_{\mathrm{cell}} = \frac{X^{/}(t)}{X(t)} + D_2 \quad (2.6.18)$$

This value appears as a solid line in Fig. 2.9 and is now a positive quantity that tends clearly toward D_2 when t tends to infinity. When t tends to infinity, (2.6.18) tends toward the value,

$$\mu|_{\mathrm{cell}} \approx D_2 \quad (2.6.19)$$

In order to take up again the kinetic relativity illustrated here in a naïve way, the following two equivalent propositions can be given,

(a) "The specific growth rate of the biomass tends toward zero in the chemostat." This expresses what is observed in the referential of the system and it is nothing other than the steady state of the system.
(b) "The specific growth rate of the cells tends toward D_2 for the new steady state." This is the result of the calculation that gives the speed in the micellian referential.

It is evident that the two propositions are true, but that the values attributed to the specific growth rates are different and depend on the referential (level of description) in which they are described.

2.7 Influence of Phase Density on Concentrations Calculation

Simple theoretical and experimental studies have, since the 1970s, shown that increase in biomass could improve productivity of continuous cultures (Monbouquette 1987). Later, it was realized that continuous techniques using high cell densities, represented a good method for avoiding cell intoxication phenomena due to release of toxic by-products. (Posten and Rinas 2000).

Walker (Walker 1998) distinguishes high cell density cultures (HCDC: *high cell density cultivation*) of which the biomass is greater than 50 g/L (dry weight, DW) and very high density cell cultures (VHCDC) that present a biomass greater than 150 gDW/L. By examining the constraints linked to such cultures (oxygenation, agitation, etc.), he quotes the example of a culture reaching DW/L that Chen et al. (1997) worked on.

So, there are production processes that use very high cell phase densities (biomass). In a representation of polyphasic dispersed systems, there is a difference between *R*- and *E*-concentrations (refer to Sect. 2.3) and this use of one or another concentration must be made sure of with particular care. However, the treatment of a problem can sometimes be very much simplified if the two concentrations are very similar and can be mixed one with another without introducing significant error (during experimentation, for example). In this section, the conditions necessary for using one or another concentration interchangeably were examined.

The experimental concentrations are, in general, *R*-concentrations. So the dose of a compound i from the culture medium, obtained by taking samples from the reactor that are measured after filtration, is expressed through a homogenous concentration obtained in the liquid phase (for example). So the quantity of compound is determined in an aliquot portion that has been made monophasic (here, by filtration). The concentration determined in the aliquot can be spread through the concentration in the corresponding phase of the system. However, it cannot be spread throughout the whole system without introducing error, if account is not taken of the density of the other phases. In fact, the working volume of the reactor is the sum of all the phase volumes. By mixing the volume of one single phase with the working volume, an error is made that in certain cases can be significant. A contrario, it is not always justifiable to devote oneself to complicated calculations to improve a value for which the experimental error is much greater than the correction made on volume.

Let us consider a concentration obtained experimentally in phase p; this is an *R*-concentration calculated by,

$$C_i^p = \frac{M_i^p}{V^p} \tag{2.7.1}$$

Moreover, the partial, pseudohomogenous concentration is given by,

$$\tilde{C}_i^p = \frac{M_i^p}{V_T} \tag{2.7.1}$$

By eliminating M_i^p from the two relationships and rearranging, it is easily found that

$$\tilde{C}_i^p = C_i^p \frac{V^p}{V_T} \tag{2.7.3}$$

Since the total volume is greater than the phase volume, it is obviously observed that the E-concentration is less than the R-concentration.

Taking into account that the total volume is the sum of all phase volumes,

$$V_T = V^p + \sum_{k \neq p} V^k \tag{2.7.4}$$

Equation (2.7.3) can be put in the form,

$$\tilde{C}_i^p = C_i^p \frac{V^p}{V^p + \sum\limits_{k \neq p} V^k} \tag{2.7.5}$$

It is clear that

$$V^p \gg \sum_{k \neq p} V^k \tag{2.7.6}$$

The two concentrations have a value that is very similar and

$$\tilde{C}_i^p \approx C_i^p \tag{2.7.7}$$

Condition (2.7.6) is the general condition that makes it possible to use either of the two concentrations by introducing just one minimum or insignificant error. So it suffices that the volume of the phase in consideration is much greater than the sum of all the other phases. This form is, however, generally very impractical and it is hoped that a less general condition that is much more helpful in practice will be introduced.

2.7.1 Two-Phase System

For this, let us consider a two-phase system: solid and liquid. The relationship (2.7.5) is reduced to,

$$\tilde{C}_i^l = C_i^l \frac{V^l}{V^l + V^s} \tag{2.7.8}$$

For the liquid phase, and the condition (2.7.6) it is reduced to,

$$V^l \gg V^s \tag{2.7.9}$$

This simply signifies that the system is very dilute and that the volume of the liquid phase is nearly equal to the working volume. Just as above, the volume of the solid phase is not generally known. Let us introduce phase density that is easily accessible. The volume of the solid phase can be expressed through the relationship of the total mass of the solid and its density,

$$V^s = \frac{M^S}{\delta_s} \tag{2.7.10}$$

By multiplying the right-hand side by V_T, and by using the definition of phase density, the following is found, (cf. (2.3.14))

$$V^s = X^s \frac{V_T}{\delta_s} \tag{2.7.11}$$

By introducing this value into (2.7.8), the following is obtained,

$$\tilde{C}_i^l = C_i^l \frac{V^l}{V^l + X^s \frac{V_T}{\delta_s}} \tag{2.7.12}$$

Taking into account that

$$V^l = V_T - V^s \tag{2.7.13}$$

Equation (2.7.8) can then be written as

$$\tilde{C}_i^l = C_i^l \frac{V_T - V^s}{V_T} \tag{2.7.14}$$

and by using (2.7.11) and by simplifying, it is found that

$$\tilde{C}_i^l = C_i^l (1 - X^s/\delta_s) \tag{2.7.15}$$

2.7.2 Numerical Example

Say, there is a reactor comprising a cell phase of phase density (biomass) 3 gDW/L (DW = dry weight). With a water content of 70 %, a wet biomass of 10 gWW/L (WW = wet weight) is obtained. The working volume of the reactor is 250 mL. By using (2.7.11), it is found that the volume of the cell phase is

$$V^c = \frac{10}{1000 \times 4} = 2.5 \times 10^{-3}\ \mathrm{L}.$$

using a cell density of 1000 g/l as above (cf. Sect. 2.2).

According to (2.7.13) the volume of the liquid phase is then

$$V^l = 0.25 - 2.5 \times 10^{-3} = 0.2475\,\mathrm{L}$$

about 100 times greater than the cell volume. Then it can be conceded that (2.7.7) is true and the *R*- or *E*-concentrations can be used interchangeably for the compound *i*. In other terms, the biomass is sufficiently small to assimilate the experimental concentration of *i* (*R*-concentration) in its pseudohomogenous concentration.

2.7.3 Critical Biomass

Let us try to estimate the limit of the biomass that can make possible this approximation. Let us represent the factor that intervenes in (2.7.15) by f_v:

$$f_V = (1 - X^c/\delta_c) \tag{2.7.16}$$

The following table gives values of f_v for different values of biomass (always supposing that the specific cell mass equals 1000 g/L and that the water content is 70 %) (Table 2.1).

In a system that can be considered as essentially made up of a liquid and a solid phase, the error is less or equal to 3 % if the biomass is less or equal to 10 gDW/L. This error can be considered as acceptable compared with experimental error over biomass measurement. Following this, it is considered that the correction is not significant for systems in which the biomass does not exceed 10 gDW/L. (So a value is obtained that is clearly lower than that conceded by Monbouquette (1987) who adopts from 20 to 25 gDW/L.) The critical biomass should be considered at present as a relatively arbitrary quantity, essentially depending on experimental measurement errors. However, it is not absurd to ask if there is another criteria (kinetic, for example) that could make this value more objective. (Such a study is not known to have been done.) For high or a very high density systems phase

Table 2.1 Variation in the correction factor between *R*- and *E*-concentrations according to biomass

X^c gDW/L	X^c gWW/L	f_v	$(1 - f_v) \cdot 100$ (%)
0	0.00	1.00	0
3	10.00	0.99	1
10	33.33	0.97	3
30	100.00	0.90	10
150	500.00	0.50	50
300	1000.00	0.00	100

density, it is estimated that the correction becomes useful, for mathematical modeling of the process, in particular. (Most often, this modeling serves for the process control.) We insist on the fact that, in the absence of a critical biomass objectification, one must keep in mind the aim of the measure and the value of experimental error (the experimental error can be very high in an industrial process), before fixing an inescapable value for the critical biomass.

2.7.4 *The Case of a Gaseous Phase*

Dispersed systems that have a gaseous phase constitute a special case. In fact, this can only be maintained in an open system according to this phase. As soon as aeration (for example) ceases, the gaseous phase disappears and the system is simplified. During a sampling, for example, the degassing is almost instantaneously done and the amount of dissolved gas gives rise to a (generally) negligible volume variation.

During aeration, on the contrary, the working volume can sometimes be considerably modified by the presence of bubbles from the gaseous phase. It is therefore necessary to be careful to be careful when defining the working volume and this is not necessarily a straightforward operation. In fact, the working volume can be defined in two ways; one consists of defining the working volume without aeration (V_T (NG)) and the other with aeration (V_T (G)). Depending on the applications, the two methods can be useful, but needless to say, only the working volume with aeration corresponds to the real situation. It is easy to note that relationship (2.7.6) can be considerably influenced by the gaseous phase, even if the solid phase is small compared to the liquid phase. If the working volume is defined with the gaseous phase, approximations defined in the preceding paragraph remain valid. Otherwise, modifications must be made. Unfortunately, in the literature, the exact characterization of the used working volume is not generally well defined. Below, when the data from the literature is used, it is supposed that the working volume given is that without the gaseous phase and the approximation of 10 gDW/L will be kept as the critical biomass before applying corrections. However, it is clear that our hypothesis does not in any way exclude the risk of some bias. Fortunately, the considered examples have biomass well below 10 gDW/L, which reduces the error range.

2.8 The Variation of the Internal Composition of a Microorganism with the Growth Rate Is a Consequence of the Mass Conservation Law

It has long been known, and a very general manner, the composition and size of a microorganism depends upon its growth rate. This dramatically applies to bacteria. We quote: "*The growth rate that a particular medium supports, not its specific composition, determines the PHYSIOLOGICAL STATE (cell size and macromolecular composition) for cells growing in it. For example, the physiological state of cells is the same in different media or even in continuous culture if the cultures are all growing at the same rates.*" (Neidhardt et al. 1994). (In a continuous system, this also applies to the growth of crystals, for example Villermaux 1982.) Here we show that the PDS formalism can easily enlighten these facts. A more detailed demonstration can be found in Thierie (1997).

The chemostat is discussed in detail in the following chapters, however, to have control over the growth rate, we still use this bioreactor as an example (see Sect. 2.6).

Consider the general mass balance Eq. (2.4.26) in the form

$$\frac{\mathrm{d}\tilde{C}_i^p}{\mathrm{d}t} = \frac{1}{V_T}\left(\tilde{F}_i^E - \tilde{F}_i^S\right) + q_i^p X^p + \tilde{C}_i^p\left(\frac{\mathrm{d}\ln N_T^p}{\mathrm{d}t} - \frac{\mathrm{d}\ln V_T}{\mathrm{d}t}\right) \tag{2.8.1}$$

This equation is valid for any compound i, in each phase p.

Let us apply this general relationship to a strictly intracellular compound i, in a constant volume chemostat. Assume also that there are no cells inflowing within the chemostat. When all these conditions are satisfied, in the steady state (2.8.1) is reduced to:

$$0 = \frac{-\tilde{F}_i^S}{V_T} + q_i^c X^c \tag{2.8.2}$$

Taking (2.4.25) into account and rearranging:

$$0 = r_i^c \frac{X^c}{\delta_c} - D\tilde{C}_i^c \tag{2.8.3}$$

Using the definition of E-concentrations to write mean concentrations, it comes after reordering that

$$0 = \left(r_i^c - DC_i^c\right)\frac{X^c}{\delta_c} \tag{2.8.4}$$

from which we obtain that finally, the steady state condition is

$$C_i^c = \frac{r_i}{D} \tag{2.8.5}$$

This relationship shows that the mean composition of a strictly intracellular compound i depends on the growth rate (D), and this irrespective of the intracellular transformation rates (r_i) of this compound. (Obviously, transformation rates can be very complex and depend on many factors. Equation (2.8.5) therefore does not imply that the internal concentration inversely varies with D.)

Mass balance, which is an implicit relationship (no "assumption" other than matter conservation) therefore provides a nontrivial and extremely general result (2.8.5).

The result seems incontestable and the steady condition (2.8.5) is a necessary condition. But is this condition sufficient? In other words, does the steady intracellular concentration condition (2.8.5) adequately fulfill both mass balance requirements and cellular effectiveness simultaneously? The question is not obvious, but we are tempted to believe that this is not the case. We know that the steady cellular growth in a chemostat is only possible in a bounded range of dilution rates. Beyond a critical value, the cellular growth fails to counterbalance the hydraulic output and the system undergoes a washout (all the cells are "washed out" of the vessel). It is possible to interpret this well-known phenomenon in terms of a conflict between the metabolic requirements for growth and the necessity to satisfy the concentration values imposed by the law of conservation of mass. The Fig. 2.10 illustrates this idea.

The curves represent the steady states of a chemostat as a function of D, for three different possible metabolic reactions (see Figure caption). For two of them (curves (1) and (2)), the ratio P/S (Product/Substrate (or reactant)) decreases asymptotically to zero but vanishes only as D tends to infinity. For finite values of D, the concentration of the product may become very small but the metabolite is always produced. Theoretically, there is no reason to expect an event to some particular value of D. On the contrary, the third mechanism (curve (3)) leads to a complete exhaustion of the metabolite at a precise value of D (around 3.2 t^{-1} in our example).

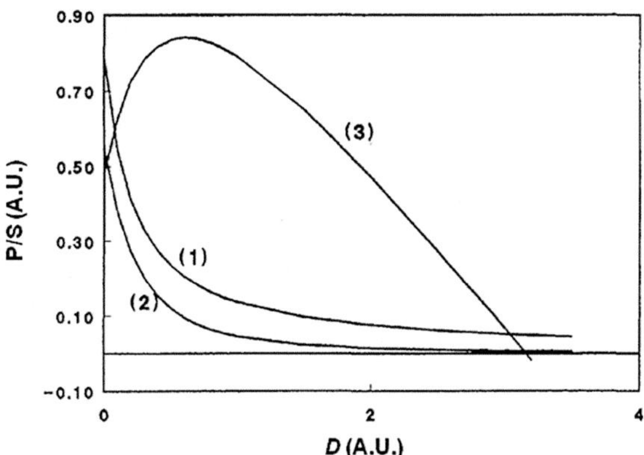

Fig. 2.10 Steady states of a chemostat as function of dilution rate, D, for different intracellular kinetics. S is a substrate (or reactant) leading to the production of P (product). The P/S ratio is plotted versus D. (*[A.U.]: arbitrary units*) Curve *1*: Monomolecular reaction: $S \xrightarrow{k} P$. Curve *2*: Bimolecular reaction: $S + S \xrightarrow{k} P$. Curve *3*: Michaelis–Menten mechanism: $S + E \overset{k+,k-}{\longleftrightarrow} (ES) \overset{k}{\longleftrightarrow} P + E$

If the metabolite is indispensable for cell growth or viability, a critical dilution rate exists beyond which the washout occurs. Experimental protocols like those depicted in Fig. 2.2 ($P/S = f(D)$) may lead to interesting information about the kinetics of two metabolites suspected to derive from each other in a more or less direct fashion. Systematic inspection of such P/S ratios as a function of dilution rate reveals specific behaviors and may be used to highlight particular links and specific physiological properties.

Without any interpretation, the concentration change due to the mass balance obscure, in some sense, the role that a metabolite may play in the growth process. Some metabolites (or organelles, like ribosomes) are supposed to play a particular and important role in the mechanism of cell growth. The correlative change of those compounds with growth rate is interpreted as governing the cellular growth. We know now that one must be very circumspect in doing so, because a change in concentration with growth may be "active," that is to say, necessary to permit this particular rate, or "passive," being a consequence of this specific rate and simply obeying the mass conservation law.

To conclude, by mean of a pseudohomogenous description of a polyphasic system, we have analyzed the consequences of the growth rate on the concentration of a strictly intracellular compound of a cell cultivated in a chemostat. Without any explicit kinetic representation, the results are kept at a qualitative level. We have chosen this example, because the effects are well known in microbiology and bioengineering and illustrate the relevance of our approach. The method, even at a qualitative level, permits to explore the influence of other parameters than the dilution rate (like external kinetics, phase number, and so on), perhaps leading to unexpected predictions and interesting challenges for experimenters. By use of explicit kinetics, our formalism must be able to provide quantitative results, allowing to conceive new perspective in modeling and different optimization and monitoring concepts.

References

Baldwin W.W. and Kubitschek H.E. (1984) Buoyant density variation during the cell cycle of *Saccharomyces cerevisiae*. J. Bacteriol. **158**(2):701-704.

Dammel E.E. and Schroeder E.D. (1991) Density of activated sludge solid. Wat. Res. **25**(7):841–846.

Dagnelie P (1981) Théorie et méthodes statistiques – Volume 1 Les Presses agronomiques de Gembloux.

Dagnelie P (1980) Théorie et méthodes statistiques – Volume 2 Les Presses agronomiques de Gembloux.

Chen Y.L., Cino J., Hart G., Freedman D., White C. and Komives E.A. (1997) High protein expression in fermentation of recombinant *Pichia pastoris* by a fed-batch process. Process Biochemistry. **32**:107–111.

Fredrickson A.G. and Tsuchiya H.M. (1963) Continuous propagation of microorganisms. A.I.Ch. E. Journal **9**(4):459–468.

Kaufman A (1965) Cours de calcul de probabilité Eds. Albin Michel.

Kubitschek H.E., Baldwin W.W. and Graetzer R (1983) Buoyant cell density constancy during the cell cycle of *Escherichia coli*. J. Bacteriol. **155**(3):1027–1032.

Kubitschek H.E., Baldwin W.W., Schroeder S.J. and Graetzer R (1984) Independence of buoyant cell density and growth rate in *Escherichia coli*. J. Bacteriol. **158**(1):296–299.

Landau L et Lifchitz E (1969) Physique théorique. Tome I. Mécanique. Ed. Mir, Moscou (3ème édition).

Minkevitch I.G. and Abramychev, Yu A (1994) The dynamics of continuous microbial culture described by cell age distribution and concentration of one substrate. Bull. Math. Biol. **56** (5):837–862.

Monbouquette H.G. (1987) Models for high cell density bioreactors must consider biomass volume fraction: Cell recycle examplc. Biotcchnol. Biocng. **29**:1075–1080.

Neidhardt F.C., Ingraham J.L. and Schaechter M (1994) Physiologie de la cellule bactérienne. Une approche moléculaire. Masson, Paris.

Posten C. and Rinas U. (2000) Control strategies for high cell densities cultivation of *Escherichia coli*. In [Schügerl and Bellgardt (2000), Chpaitre 12).

Roels J.A. (1983) Energetics and Kinetics in Biotechnology. Elsevier Biomedical Press, Amsterdam.

Schügerl K and Bellgardt K.H. (Eds) (2000) Bioreaction Engineering – Modeling and Control. Springer-Verlag, Berlin, Heidelberg.

Takeo M (1999) Disperse systems Wiley-VCH, Weinhein (Federal Republic of Germany).

Thierie J (1997) Why does bacterial composition change with the chémostat dilution rate? Biotechnology Technique **11:**625–629.

Thierie J, Bensaid A and Penninckx M (1999) Robustness, coherence and complex behaviors of a bacterial consortium from an activated sludge cultivated in a chemostat. Thirteenth Forum for applied Biotechnology. Med. Fac. Landbouw. Univ. Gent **64/5a**:205–210.

Tsuchiya H.G., Fredrickson A.H. and Aris R. (1966) Dynamics of microbial cell populations. Adv. Chem. Engineering **6**:125–206.

Villermaux J (1982) Génie de la réaction chimique : conception et fonctionnement des réacteurs. Technique et Documentation (Lavoisier), Paris.

Walker G.M. (1998) Yeast – Physiology and Biotechnology. John Wiley & Sons, Chichester, England.

Wolfe A. J. (2005) The acetate switch. Microbiology and molecular biology reviews, **69**:12–50.

Chapter 3
Continuous Culture: The Chemostat

Abstract This chapter is the heart of the book. Its purpose is twofold: first, it serves to demonstrate the effectiveness of representation in polyphasic dispersed systems (PDS); second, it illustrates, with examples, effectiveness and originality of the method. The cases are first treated in an implicit form, that is to say only using the raw form of mass balance, without any concept other than the matter conservation. Besides the simple cases at steady state (including the problem of water), we treat several cases of bacterial synthesis (cytochromes, proteins, RNA, etc.). But the most important "implicit" result is probably that the concept of "Maintenance Energy", where we show that the identification of cryptic flow in the PDS enables us to offer an explanation of wasted energy in the form of substrate recycling. The most remarkable result is probably obtained using the mass balances under their explicit form (by "modeling"). We clearly show that the Crabtree effect in *Saccharomyces cerevisiae* is due to an overflow of the oxidative metabolic pathway resulting from the existence of at least two transport pathways: one with low affinity and the other with high affinity for the substrate. We show that this model can very unexpectedly be extended to more complex cases, such as respiro-fermentative transitions in continuous cultures of floc-forming bacteria.

3.1 General Remarks

Usually a distinction of thermodynamic (classic or otherwise) is made of isolated, closed, or open systems (Prigogine 1968; Nicolis and Prigogine 1977). The first of these exchanges neither energy nor matter with the outside world; the second exchanges only energy; and the third exchanges energy and matter with their environment.

From this strict point of view of a physicist, all cell cultures are open systems. The most simple liquid culture system is the "batch." Appropriate nutrients are placed in a container, an inoculum is added and the rest is up to nature (the principle is the same in a solid culture, like on a petri dish). This biotechnological device is

J. Thierie, *Introduction to Polyphasic Dispersed Systems Theory*,
DOI 10.1007/978-3-319-27853-7_3

several thousand years old (as in the production of beer by the Egyptians of antiquity) and is also used by present day by populations "called primitive" (for various alcoholic fermentations, etc.). It is always current, and doubtless one of the most used devices in our modern laboratories.

It will clearly appear as a nonisolated system, because the batch is usually either thermally nonisolated or thermostated. In both cases, it exchanges heat with the environment. In fact, most of the reactions of the cell metabolism are usually exothermic and there is production of heat in the core of the culture.

On the other hand, the batch does not seem to be an open system intuitively. Effectively, microorganisms and culture medium are confined in the container and solid and liquid phases are effectively separated from the outside world. However, whether the culture is aerobic or anaerobic, there is a gaseous flow between the culture and the outside world. In the case of an aerobic culture, air (or oxygen) is provided to the system and a different gas mixture (enriched with carbon dioxide, ethanol vapor, for example) leaves it. In the case of anaerobic cultures, there is at least always CO_2 leaving the cell culture. (Other gases and/or volatile substances can also escape from the system.)

It should also be admitted that strictly speaking, all cultures (in any case all the usual cultures) are in reality, open systems. However, there is a distinction between the methods of culture that are only open in relationship to the gaseous phase and those which exchange matter in other physical states (liquids and/or suspended solids). These latter belong to the category of continuous or semi-continuous cultures (that are both called continuous cultures to make the text easier).

Historically, continuous cultures are far more recent. It has been seen that batch cultures have a long traditional history and certain among them are several centuries old.

Continuous cultures are the most recent and we think that their young age (scarcely more than 50 years) is one of the reasons that affects the numbers of them introduced into current usage. (It is certainly not the only reason, but the discussion of the pros and cons of different types of culture is not the subject here. We would just point up that the encouragement of some types of continuous culture remains to be completed.)

3.1.1 Historical Overview

When the general treatises on microbiology or biotechnology are consulted, it seems evident that it is necessary to begin continuous cultures (and especially the chemostat) with Monod (the "bactogene"; Monod 1949, 1950) and Novick and Szilard (1950). 1950 is effectively and without doubt, the date that marks the infamous advent of these new types of processes. Panikov (1995), however, has noted that the concept is without doubt most ancient and traces back the technique of Winogradsky in 1890. This microbiologist worked in Strasbourg and used to study the growth of sulfurous bacteria under the microscope. He fed daily the observed droplet with a solution of hydrogen sulfide. By doing this, the culture certainly was discontinuous, but gave nevertheless rise to a flow with the outside world. Between 1890 and 1950, other discontinuous cultures were used, notably by Woodruff and Baitsell (1911a, b) or by Galadzhiev (1932) for the culture of protozoa. The true continuous feeding methods made their advent between 1920 and 1950 and are used for fermentations (Panikov 1995).

3.1.2 The New Paradigm

On the basis of the first review concerning continuous cultures (McClung 1949), Panikov (1995) defends the following double thesis:

1. until 1949, continuous cultures were only thought of as techniques used to refresh cultures, whether by adding fresh nutrients, or by eliminating exhausted compounds. So, there is no perception of the innovative properties of these types of cultures.
2. awareness of particular features of continuous cultures (the new paradigm) is due to Monod, because he was capable of giving a theoretical representation of the phenomenon.

This analysis is both provoking and exciting.

Provoking, because it implies that a new level of understanding (the characteristics belonging to a new technique) must sometimes have a theoretical representation of the phenomenon to be fully understood. This assertion underlines the major importance of the (mathematical) modeling of biological systems (in this case).

Exciting, because the theoretical representation of a complex process, such as continuous culture, is always a challenge. There is no recipe to follow, particularly when it is a question of constructing a model. Several people who construct models have not hesitated to declare the modeling is more like an art than an objective method (Segel 1984). I permit myself the pleasure of quoting Skellam (1973) (as he appears in Brown and Rothery 1993): "*Roughly speaking, a model is a peculiar blend of fact and fantasy, of truth, half-truth and falsehood. In some ways a model may be reliable, in other ways only helpful and at times and in some respects thoroughly misleading. The fashionable dogma that hypothetical schemes can be tested in their totality in some absolute sense, is hardly conducive to creative thinking. It is, indeed, just as great a mistake to take the imperfections of our models too seriously as it is to ignore them altogether. ...*". Certainly the subject dates from 1973 and can be called strict in view of the progress of modern modeling. Nevertheless, in many fields, Skellam's ironic description preserves a freshness and a currency that doubtless will never go out of fashion. It is therefore particularly interesting to construct steps that make possible guidance of the construction of a model in a way that makes it tend as rationally as possible toward a functional understanding of the reality.

Note As seen in Chap. 2, until now PDS **are not mathematical models**, but simply a particular way of looking at a system and writing its mass balances. Modeling strictly speaking will be considered later.

3.2 The Chemostat and Monod's Model

3.2.1 The Beginnings

Again, it is from Panikov (1995) that the broad themes of this section are borrowed. This author makes very clear the development of the theory of the chemostat for the development of continuous culture techniques. Notably, he demonstrates the difficulties in bringing in the essential newness of the chemostat. He draws on a significant anecdote to illustrate his subject. As will be seen in the following section, the chemostat can reach a series of steady states where cell growth is equal to the volumetric flux of substrate per unit of useful volume. On reflection, this result is quite extraordinary since it implies that several thousand metabolic reactions adjust simultaneously to make possible both cell viability and a precise growth rate that is imposed by the experimenter. It seems that this fact has been extremely difficult to understand for a series of commentators on Monod's work. So it can be read in Golle (1953), "*there is only one rate of medium flow... at which steady-state conditions will be maintained*," a conclusion that Panikov rated as, "ridiculous." It must be noted that the article dates from 1953, from which it might be said that a right apprehension of the chemostat by specialists took several years. It is thought that this fact could also have delayed the penetration of the concepts of continuous cultures into the heart of certain scientific and industrial milieux.

However, very soon after Monod's work, Novick and Szilard; several schools formed to study these new concepts (Portdown, Prague…).

In spite of certain "slipups," continuous cultures (and semi-continuous) owe a great deal of their success to the opportunities for application of these processes to industrial production. Since the 1950s, production of ethanol and baker's yeast has been inspired by these new techniques; the 1960s saw the development of mass production of *single-cell protein* (SCP) (Panikov 1995). Numerous variations on the theme of the chemostat are developed to achieve these objectives, both at the level of processes and at the level of theories. It is thought here, contrary to what Panikov seems to confirm (and he does not cite references on the subject), that it was above all the semi-continuous techniques that developed in industry, on a big scale. Cultures that are really continuous, such as the chemostat and its variations (recycled, cascade, forced, etc.) are subject to prejudice from industrial producers. It is thought that the relative newness of the procedures and false interpretations from the beginning are partly the cause of this. This is one of the reasons that compelled the choice of the chemostat as a preferred subject for the application of PDS. Recent experience with the baker's yeast production industry encourages the thought that the battle is far from won, and that industrialists who take up the challenge of undertaking (again) the adventure of continuous production in new domains, are perhaps, not yet born.

3.2.2 Summary Presentation of the Chemostat

Figure 3.1 represents schematically, the structure of the bioreactor called the "chemostat." Essentially, it consists of a vat [1] (glass or stainless steel according to

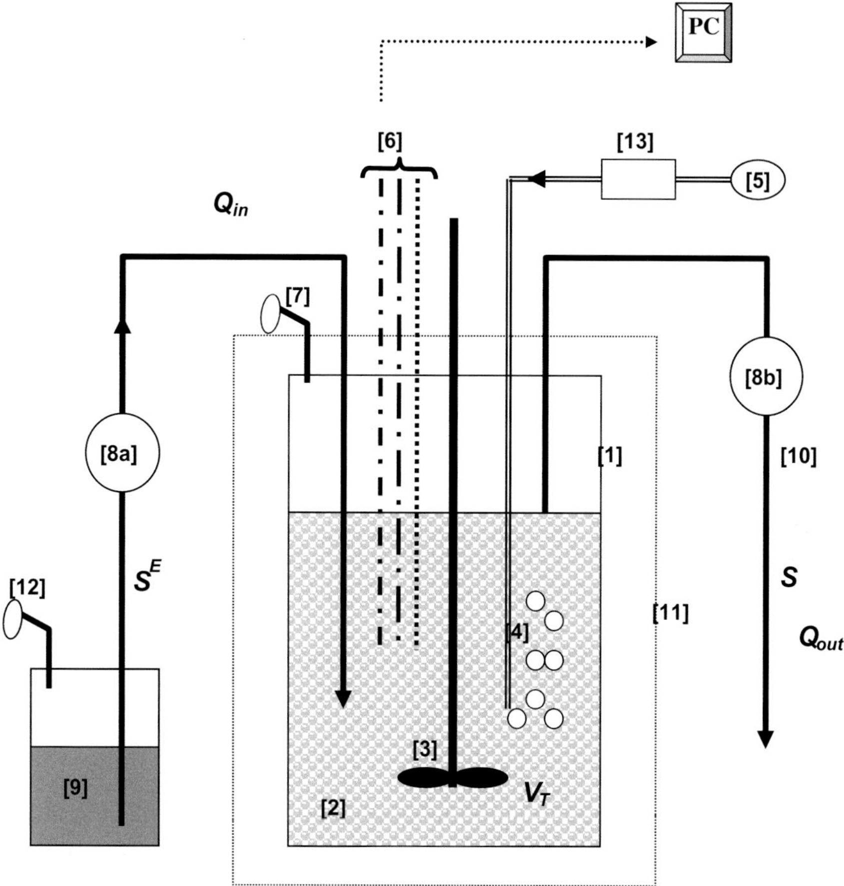

Fig. 3.1 Simplified diagram of a chemostat. [*1*] Bioreactor wall, [*2*] culture medium and cells, [*3*] stirrer, [*4*] gas diffuser (air, oxygen, nitrogen, etc.), [*5*] blower, [*6*] various sensors (t°, pH, O_2...), [*7*] sterile outlet for gas or measuring devices (CO_2...), [*8a, b*] pump and volume flowmeter, [*9*] medium of sterile culture, [*10*] outlet for excess medium, [*11*] thermostat, [*12*] porous sterilizing plug (0.2 μm), [*13*] gas flowmeter and humidifier. Q_{in} and Q_{out} are the volume inflow and outflow, respectively; V_T is the total useful volume and S_0 and S are the concentrations of limiting substrate at the inlet and outlet of the chemostat

size), usually thermostatic [11]. On one side is the nutritional medium [9] at concentration S_0, characterized by a limiting substrate (nutrient), is brought into the vat by a pump [8a] with a flow volume Q_{in} (in volume per unit of time). Moreover, the excess fluid from the bioreactor is eliminated, often by another pump [8b], so that the useful volume V_T is kept constant, and this implies that the outflow Q_{out} equals the inflow $Q_{in} = Q_{out} = Q$. Taking into account the cell consumption of the limiting substrate in the outlet [10] is S.

An agitation system [3] makes certain as possible the perfect mix at the center of the bioreactor. (This agitation system can differ greatly according to the type of culture (prokaryotes, yeasts, mammalian cells, etc.)) A blower [5] makes it possible to send gas into the culture (oxygen or air, for aerobic cultures; nitrogen of neutral gas for anaerobic cultures, etc.). A rotameter (or any gas flux measuring system) [13] and/or a humidification system (bubbler) is placed between the blower and the gas diffuser [4] submerged in the culture medium. A porous plug device [7] with weak porosity (0.22 μm, for example) makes it possible to avoid a rise in pressure in the chemostat while making sure that it is sterile. (A similar device [12] is placed on the sterile reservoir of substrate, this time to avoid lowering pressure in the vat.) Finally, a series of sensors [6] makes possible online (or other) measurement of various parameters such as the temperature, pH, redox potential, and sometimes even the biomass.

A comment is necessary for the determination of useful volume V_T. This quantity is extremely important, notably for the precise determination of the dilution rate (refer to the following section). Now, the estimation of this volume is much more difficult than was at first thought. It seems that, in the literature, the useful volume is generally confused with the nominal volume of the reactor, that is to say with the volume of the reactor when empty (possibly such as that defined by the manufacturer). For aerobic cultures (or including a gaseous phase, in general), the volume with and without gas bubbles can sometimes vary in a very spectacular manner. As far as high density cell cultures are concerned, the volume of the liquid phase alone can differ significantly from the liquid + solid volume, etc. There is no clear answer at the present time, as to what the definition of useful volume should be. It is evident that its assimilation with the nominal volume can lead to errors (at the time of comparisons with interlaboratory experiments, for example). It is possible that a definition that is more or less arbitrary for this volume must suffice, since its definition cannot be made objective. When the chemostat was used in the laboratory (Thierie et al. 1996; Bensaid et al. 2000; Thierie and Penninckx 2004; Thierie et al. 1999), the useful volume was always defined as the sum of the liquid + solid phases (cells or bacterial flocs). Given that the biomass was relatively low (less than 10 gDW/L), it was possible to estimate that the useful volume was largely assimilated at the liquid phase.

3.2.3 *Monod's Model*

It is nearly impossible to talk about the chemostat without presenting Monod's model. No delays will be made by going into the laws of evolution of the model that are abundantly presented elsewhere in their general forms (Monod 1950; Doran 1995) or presented in certain, particular mathematical aspects (Segel 1984; Waltman 1983). However, a little time will be spent on the stability (in the sense used by Lyapounov (Kuznetsov 1995) of the model, by presenting a complete analysis that has not been found elsewhere. There are reservations made over the relevance of Monod's model and an analysis of this question has been completed but is not given here so that the text is not overloaded. For the presentation of the model, the usual formulae have been adopted and not those of PDS in order to emphasize that the steps adopted in the two methods of presentation are essentially different.

Generally, the model is presented in this form,

$$\frac{\mathrm{d}X}{\mathrm{d}t} = X(\mu(S) - D) \tag{3.2.1}$$

$$\frac{\mathrm{d}S}{\mathrm{d}t} = D(S^E - S) - \mu(S)X/Y \tag{3.2.2}$$

So it is a question of a determining model with two variables in state, X, the biomass (or the cell density, in dry weight (DW) mass per unit of volume), and S, the limiting substrate (also in mass per unit of volume). S^E is the concentration at the inlet of the chemostat (cf. Fig. 3.1). D is the dilution rate, that is to say the ratio Q/V_T. Finally, Y is the yield coefficient, assumed constant in Monod's model and of which the definition is given by,

$$Y = \frac{X}{S^E - S}$$

This measures the quantity of biomass produced per unit of limiting substrate consumed.

The specific growth rate is presumed to depend only on the limiting substrate and Monod chose the following hyperbolic form:

$$\mu(S) = \mu_{\max}\frac{S}{K_M + S} \tag{3.2.3}$$

where $\mu_{\max}$ is the maximum specific growth rate (constant) and K_M is Monod's constant. (This last quantity is often considered as the inverse of infinity of the rate in relationship to the substrate, probably because of the analogy between (3.2.3) enzymatic kinetics of Michaelis–Menten.) All of the relationships (3.2.1–3.2.3) define Monod's model.

In practice, the chemostat is started off by filling the vat [1] to half full with the culture medium and an inoculum is added (a dose) of microorganisms. The culture is left in this state for from a few hours to several days (according to the growth rate of the microorganisms), following which the pumps [8a, b] are set into action in such a way that a flux Q occurs that defines the dilution D rate. The system starts up in batch before moving on to chemostat mode (for various technical reasons (Demain and Solomon 1986)). The bioreactor functions from then on without modification until the state variables (and other parameters) take on constant values. While the system is evolving, this is called a transient regime; when the variables become constant, the system is said to be in a steady state.

(This is not a state of equilibrium in the thermodynamic sense, since the system requires energy to remain in this state; some is dissipated too (heat due to growth of microorganisms, thermic losses from pumps, etc.) and so it is a dissipating system). It is well known that Monod's model does not really take into account the transient regime (Panikov 1995; Roques et al. 1982). So Monod's model of the chemostat is used for these steady states. Since the variables of state are constant in the steady state, the condition of being stationary is written in a general manner as the cancelation of the derivatives,

$$\frac{\mathrm{d}Z_i}{\mathrm{d}t} = 0; \quad \forall i$$

If this condition is applied to (3.2.1), it is seen that two solutions are possible,

$$\mu(S) = D \tag{3.2.4}$$

$$X = 0 \tag{3.2.5}$$

The first steady state implies that the specific growth rate of the cells adjusts in a way that compensates for the outflow of the biomass. For ease, this state will be called the nontrivial state. The second steady state will be called trivial because it implies very simply that the chemostat is empty of all biomass.

By injecting (3.2.5) into (3.2.2), it is easy to see that this trivial state of biomass corresponds to the following state of the substrate:

$$D(S^E - S) = 0$$

from which it is easily deduced that

$$S = S^E \tag{3.2.6}$$

In the absence of biomass, there is no consumption of substrate, so the concentration remains equal to the concentration at the inlet.

To find the steady state of the substrate that corresponds to the nontrivial state, it suffices to explain the relationship (3.3.4) due to (3.2.3), so,

$$\mu_{\max}\frac{S}{K_M + S} = D$$

from which it can be shown that

$$S = \frac{DK_M}{\mu_{\max} - D} \tag{3.2.7}$$

(Note that $\mu_{\max} = D$ is a singular point.)

Under condition (3.2.4), the relationship (3.2.2) is put in the form,

$$S^E - S = X/Y$$

By rearranging,

$$X = Y(S^E - S) \tag{3.2.8}$$

which gives the value of biomass in the nontrivial steady state. (It is interesting to note that this state corresponds exactly to the definition of yield coefficient given above.)

It can be noted that, in this model, the steady states are independent of the size of the inoculum, that is to say the initial condition of the biomass. In retrospect, the nontrivial state depends on the concentration of the substrate at the inlet.

3.2.4 *Stability of Steady States*

It has just been seen that Monod's model has two steady states, that is to say, multiple steady states. It is evident that a system cannot be in two states at once. What dictates the state of a system that can have multiple states is the stability of the state considered and the situation of the system on the path in the phases' space. The situation here (two states) is quite simple.

The stability of a steady state can be determined by an analysis of the normal modes. The method is described elsewhere (Nicolis and Prigogine 1977). To simplify matters, let us say that the linearization of a system makes it possible to examine the behavior of pathways in the vicinity of a steady state. On one dimension, it could be written that in the vicinity of the steady state Z^0, the trajectory would be of the type,

$$Z(t) = Z^0 \exp(\omega t)$$

Let us assume that ω is real. If $\omega < 0$, then the system would regain its steady state if a disturbance (moderately) separated it from it. The state is then qualified as stable. Inversely, for $\omega > 0$, it holds that $Z(t) \rightarrow \infty$ and all disturbances (even infinitesimal) of the steady state would cause the system to move away from it. The steady state is then qualified as unstable. In the case where $\omega = 0$, the system is called marginally stable. It is enough to know that a system is stable if all normal modes are negative. Practically speaking, the analysis of normal modes can be achieved by resolving the equation that represents the system (in matrix notation).

$$\text{dét}|\mathbf{J} - \omega\mathbf{U}| = 0 \tag{3.2.9}$$

where $\mathbf{J}$ is the Jacobian of the system (3.2.1–3.2.2) and $\mathbf{U}$ the unit matrix (Kuznetsov 1995). The values of the state variables are taken at the steady state.

In an explicit form, (3.2.9) takes the form,

$$\begin{vmatrix} \mu(s) - D - \omega & X\partial_S\mu(S) \\ -\mu(S)/Y & -D - X/Y\partial_S\mu(S) - \omega \end{vmatrix} = 0 \tag{3.2.10}$$

where

$$\partial_S\mu(S) = \frac{K_M\mu_{\max}}{(K_M + S)^2} > 0 \tag{3.2.11}$$

Trivial case For the trivial state (cf. Table 3.1), (3.2.10) is simplified,

$$\begin{vmatrix} \mu(s) - D - \omega & 0 \\ -\mu(S)/Y & -(D + \omega) \end{vmatrix} = 0$$

and is put in the form,

$$(D - \mu(S) + \omega)(D + \omega) = 0 \tag{3.2.12}$$

that allows the solutions,

$$\omega_1 = -D \tag{3.2.13a}$$

$$\omega_2 = \mu(S) - D \tag{3.2.13b}$$

The first solution is always negative, since D is a positive quantity (the case $D = 0$ is a special case). The second solution is negative only if

$$D > \mu(S)$$

Table 3.1 Steady states of Monod's model

	Biomass	Substrate	Relationship
S.S. trivial	$X = 0$	$S = S^E$	X (3.2.5) S (3.2.6)
S.S. nontrivial	$X = Y(S^E - S)$	$S = \frac{DK_M}{\mu_{\max} - D}$	X (3.2.8) S (3.2.7)

The critical case is that for which $\omega_2 = 0$, that is to say, as $S = S^E$

$$D_W = \frac{\mu_{\max} S^E}{K_M + S^E} \tag{3.2.14}$$

The analysis of stability shows therefore that the trivial state only becomes stable when the dilution rate reaches the critical value, D_W, slightly less than $\mu_{\max}$. (The singular point of (3.2.7) is therefore never reached.) So, when the substrate flux becomes too big, all the biomass is removed from the chemostat; this phenomenon is known as the *washout*.

Nontrivial case In this case, the general relationship (3.2.10) takes the following form:

$$\begin{vmatrix} -\omega & X\partial_S\mu(S) \\ -\mu(S)/Y & -D - X/Y\partial_S\mu(S) - \omega \end{vmatrix} = 0 \tag{3.2.15}$$

By putting $h = X/Y\partial_S\mu(S)$ and developing, it is found that

$$\omega^2 + \omega(D+h) + \mu(s)h = 0$$

for which the solutions are,

$$\omega_{1,2} = \frac{-(D+h) \pm (D-h)}{2}$$

and it is found that

$$\begin{aligned} \omega_1 &= -D \\ \omega_2 &= -h \end{aligned} \tag{3.2.16a}$$

or

$$\omega_2 = -X/Y\partial_S\mu(S) \tag{3.2.16b}$$

Again, the first solution is always negative. The second solution is always negative, too, except when $X = 0$, and this corresponds to the trivial case. In reality, ω_2 can become positive if X becomes negative. Obviously, this is not a solution that is biologically accceptable, but this calculation will help us understand better what is happening.

Using (3.2.8) and (3.2.7), it is found that $X(D)$ has the following form:

$$X = Y\left(S^E - \frac{K_M D}{\mu_{\max} - D}\right) \tag{3.2.17}$$

It is clear that $X(D)$ is a decreasing function such that $D < \mu_{max}$. The critical condition intervenes when $X = 0$, that is to sayusing (3.2.17) when

$$S^E = \frac{K_M D}{\mu_{max} - D} \tag{3.2.18}$$

From which it is easily found that the value D corresponds to $X = 0$.
So,

$$D(X = 0) = \frac{\mu_{max} S^E}{K_M + S^E} \tag{3.2.19}$$

This value is exactly equal to (3.2.14) and corresponds to the critical dilution rate for which the trivial state becomes stable.

To summarize, when $X > 0$, the nontrivial state is stable and when $X < 0$, the trivial state is stable. The relationship (3.2.19) corresponds to marginal stability ($\omega_2 = 0$) of the nontrivial state. In other terms, $D = D_W$ corresponds exactly to the change in stability of the states, so the nontrivial state becomes unstable when the trivial state becomes stable. This phenomenon is known by the name of "exchange of stabilities" (Chandrasekhar 1970; Glansdorf and Prigogine 1971).

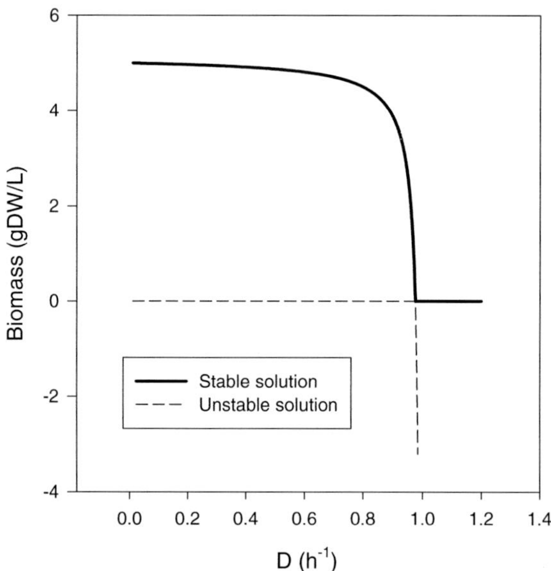

Fig. 3.2 Steady states of biomass according to dilution rate. Stable solutions are represented by *solid lines*; unstable solutions, by *dashes*. The solution $X = 0$ ceases to be unstable exactly when $D = D_W$, and becomes stable beyond this value. The stability of $X > 0$ follows exactly the reverse behavior. So when $D = D_W$, the solutions "exchange their stability." The parameters used for the simulation are, $\mu_{max} = 1\ h^{-1}$; $Y = 0.5$; $K_M = 0.25$ g/L; and $S^0 = 10$ g/L. The critical value is $D_W = 0.9760\ h^{-1}$

So, in spite of its apparent simplicity, Monod's model already presents interesting, dynamic properties. Figures 3.2 and 3.3 show the stable branches and the unstable branches of the model's two solutions.

What is interesting is that the big difference between Monod's representation and the representation of PDS is obviously the absence (explicit) of different phases in Monod's model. So there is no formal representation of interphasic exchange flux. The second term of (3.2.2) is therefore a consumption rate of the limiting substrate by the cells and not an interphasic exchange flux. This is an extremely important, different approach because the definition of the consumption rate presupposes a hypothesis of the mechanism. In Monod's model, the substrate consumption rate is proportional to the total growth rate, μX, the proportionality coefficient being $1/Y$ that is considered constant. It is a question of a model (of a hypothesis) and not of a balance. The best proof of this is that, very quickly the yield coefficient is shown not to be a constant and this has led to a wealth of adaptation or modifications of the model, often ad hoc. This problem will be looked at in detail in the section devoted to the concept of maintenance. What interests us now is how to describe the chemostat in PDS formalism.

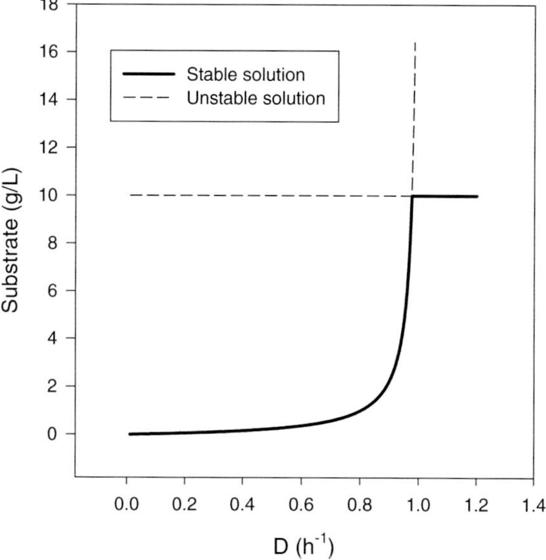

Fig. 3.3 Steady states of the substrate according to the dilution rate. The stable solutions are represented by *solid lines*; the unstable solutions, by *dashes*. Solution S = 10 g/l ceases to be unstable exactly when $D = D_W$, and becomes stable beyond this value. The stability of $S < 10$ g/L behaves in exactly the opposite way. So, when $D = D_W$, the solutions "exchange their stability." The parameters used for the simulation are $\mu_{max} = 1\ h^{-1}$; $Y = 0.5$; $K_M = 0.25$ g/L; and $S^0 = 10$ g/L. The critical value is $D_W = 0.9760\ h^{-1}$

3.3 The Chemostat in PDS Theory

3.3.1 The General Mass Balance

Let us look again at the general balance (2.4.26) obtained in the preceding chapter:

$$\frac{\mathrm{d}\tilde{C}_i^p}{\mathrm{d}t} = \frac{\tilde{F}_i^{p,E} - \tilde{F}_i^{p,S}}{V_T} + q_i^p X^p + \Phi^0_{i,\{q \neq p\}}(p) + \tilde{C}_i^p \left(\frac{\mathrm{d} \ln N_T^p}{\mathrm{d}t} - \frac{\mathrm{d} \ln V_T}{\mathrm{d}t} \right) \quad (3.3.1)$$

and let us represent the chemostat in the following, very simplified form.

Whatever is the definition of the useful volume of the chemostat (refer to remarks in the preceding section), the useful volume must be constant and so,

$$\frac{\mathrm{d} \ln V_T}{\mathrm{d}t} = 0 \quad (3.3.2)$$

In a general manner, the flows are expressed in the following manner (Villermaux 1982) in the situation that appears in Fig. 3.4:

$$\tilde{F}_i^{p,Z} = Q^Z \times \tilde{C}_i^{p,Z} \quad (3.3.3)$$

where Z indicates the locality of the flow (Inflow, Outflow, etc.).

It is important to mention that the flow must be expressed in terms of E-concentrations. In fact, it is rare that the inflow is polyphasic, it is always at least biphasic at the exit (micelle, residues from the culture medium, secondary metabolites, etc.). The relationship (3.3.3) is then general and applies to the exit as well as elsewhere (Thierie 1997). It is without doubt less usual in the case of a

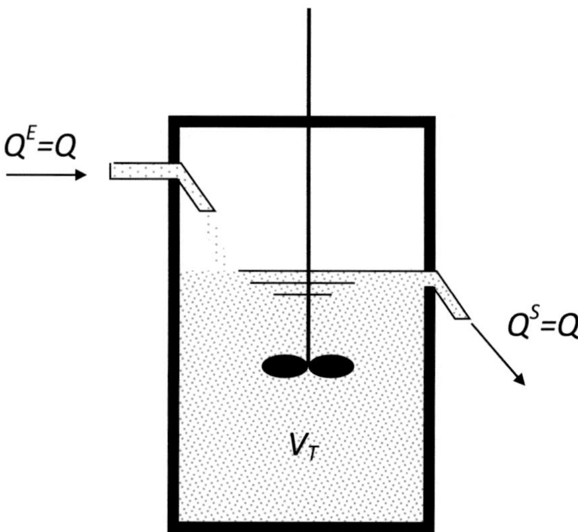

Fig. 3.4 A very simplified representation of the chemostat

gaseous phase, but can also be applied to this; it even presents a certain interest in polyphasic cases (of the aerosol type or solid particles in suspension) but this eventuality will not be studied in this work.

Taking into account (3.3.2) and the fact that $Q^E = Q^S = Q$, the relationship (3.3.1) takes the form

$$\frac{\mathrm{d}\tilde{C}_i^p}{\mathrm{d}t} = \frac{Q}{V_T}\left(\tilde{C}_i^{p,E} - \tilde{C}_i^{p,S}\right) + q_i^p X^p + \Phi^0_{i,\{q\neq p\}}(p) + \tilde{C}_i^p \frac{\mathrm{d}\ln N_T^p}{\mathrm{d}t} \tag{3.3.4}$$

The dilution rate D, is now defined, as the ratio of the volumetric flow to the useful volume:

$$D = \frac{Q}{V_T} \tag{3.3.5}$$

This quantity has dimensions that are the inverse of a time (t^{-1}). The inverse of the dilution rate is the retention time:

$$\frac{1}{D} = \tau = \frac{V_T}{Q} \tag{3.3.6}$$

and obviously has the dimensions of a time (t).

Using the definition of the dilution rate in (3.3.4), the required balance is finally seen to be:

$$\frac{\mathrm{d}\tilde{C}_i^p}{\mathrm{d}t} = D\left(\tilde{C}_i^{p,E} - \tilde{C}_i^{p,S}\right) + q_i^p X^p + \Phi^0_{i,\{q\neq p\}}(p) + \tilde{C}_i^p \frac{\mathrm{d}\ln N_T^p}{\mathrm{d}t} \tag{3.3.7}$$

This form is the general form of the mass balance for a product i of phase p of the chemostat.

Given the exchange terms with the outside world, it, however, does not apply to inputs/outputs of gases from the reactor. So, in this relationship, the exchange terms with the outside world are defined as explicit, whereas this is not the case in (3.3.1) where all the terms are implicit. Equation (3.3.7) can be called a "semi-explicit" form of the chemostat. However, only phenomena occurring in a chemostat will be studied, the term "implicit" will be kept for relationship (3.3.7). The theory will therefore be stated as:

- implicit if the exchange terms with the outside world alone are defined in an explicit way;
- explicit if at least one of the other terms is put in explicit form, that is to say, expressed in terms of state variables, kinetic constants, various physicochemical parameters, etc.

For example, the form q_i^p denoting the specific rate is implicit, whereas $q_i^p = \left(\left\{C_j^k\right\}, \{\lambda_r\}, P, T, \ldots\right)$ is an explicit form.

It must be recalled that the implicit form (3.3.7) is the expression of a mass balance and does not rest on the hypothesis of the conservation of mass. An explicit form presupposes one (or several) hypotheses for its expression, in terms of variables, constants, etc. The explicit form is therefore a form that can be shown as a mathematical model, with all the advantages and inconveniences that this entails.

3.3.2 The Biphasic Chemostat

The relationship (3.3.7) is still very general and can be applied to an indefinite number of compounds that belong to phases that enter and/or leave by the ways represented in Fig. 3.4.

For example, several nonmiscible, liquid phases that are well mixed could be imagined. In the same way, several species of cells, debris from cells or compounds seen in suspension, might form several solid phases. All these cases are possible, but certain among them are very improbable in practice.

Let us consider the simplest case, namely that of the biphasic chemostat that is to say that it is formed of only two phases—one solid and the other liquid. It is not certain that this is the most realistic situation but it proves that this theory has been already very effective in describing a great number of phenomena, possibly even already complex phenomena. The biphasic chemostat can be considered as an efficient idealization of real situations.

To represent work conditions used in the biphasic chemostat, the following situation will be adopted:

The biphasic chemostat is made up of only one cell phase (denoted by a superscript *c*) and one liquid phase (the culture medium in the broad sense), called matrix (indicated by a high *m*). The usual working conditions of the chemostat are as follows:

- the useful volume V_T, is constant;
- the volumetric flow at the entry is constant and equal to the outflow $Q^E = Q^S = Q$;
- the concentration of the compound *S* at the entry $\tilde{C}_S^{m,E}$, is constant;
- the cell phase is made up of just one species of cell (pure strain);
- the cell viability is near to 100 %;
- there are no cells in the influent: $\tilde{C}_i^{c,E} = 0$;
- the concentration at the exit is equal to that in the body of the reactor: $\tilde{C}_i^{p,S} = \tilde{C}_i^p$; $p = c, m$;

To these usual conditions, are added the following conditions:

- the average specific cell mass, δ_c, is independent of the growth rate;
- the mean water content of the cells is constant and independent of the growth rate.

(These last two conditions will be discussed elsewhere.)

For the rest, all quantities represented by a mass per unit of volume and per unit of time and specific rate, the same value brought to the unit of biomass will be called "flux." A specific rate is therefore a flux per unit of biomass (units: h^{-1}, for example).

In a system made up of two phases c and m alone, the interphasic exchange flow is reduced to

$$\Phi^0_{i,\{q\neq p\}}(p) = \Phi^0_{i,c}(m) \tag{3.3.8}$$

With two phases, obviously there is only one flux and so necessarily,

$$\Phi^0_{i,c}(m) = -\Phi^0_{i,m}(c) \tag{3.3.9}$$

This relationship simply expresses that all compounds i leaving a phase are found again in another.

Using the balance (3.3.7), the whole system belonging to the laws of evolution corresponding to the biphasic chemostat is generally written as,

$$\begin{cases} \dfrac{\mathrm{d}\tilde{C}^c_i}{\mathrm{d}t} = -D\tilde{C}^c_i + q^c_i X^c + \Phi^0_{i,m}(c) + \tilde{C}^c_i \dfrac{\mathrm{d}\ln N^c_T}{\mathrm{d}t} \\ \dfrac{\mathrm{d}\tilde{C}^m_i}{\mathrm{d}t} = D\left(\tilde{C}^{m,E}_i - \tilde{C}^m_i\right) + q^m_i X^m + \Phi^0_{i,c}(m) + \tilde{C}^m_i \dfrac{\mathrm{d}\ln N^m_T}{\mathrm{d}t} \end{cases} \tag{3.3.10a, b}$$

This equation is general and applies in all cases corresponding to the biphasic chemostat. In what follows consideration will be given to dilute systems (refer to Sect. 2.1) for which the liquid phase is a dispersing phase and does not produce micelles. In these conditions, the term "variational" in (3.3.10b) is canceled. In a dispersing phase, the expression for rate in terms of micelles no longer makes sense and a more usual formalism is returned with reference to the kinetics and it is written as,

$$q^m_i X^m \equiv R^m_i(.) \tag{3.3.11}$$

where $R^m_i(.)$ expresses a rate resulting from the transformation of a product i in the matrix phase. (Notation "(.)" indicates that this reaction is possibly complex and depends on several variables, parameters, constants, etc.)

The system that describes the evolution of the biphasic chemostat with dilute medium then takes the following form:

$$\begin{cases} \dfrac{\mathrm{d}\tilde{C}^c_i}{\mathrm{d}t} = -D\tilde{C}^c_i + q^c_i X^c + \Phi^0_{i,m}(c) + \tilde{C}^c_i \dfrac{\mathrm{d}\ln N^c_T}{\mathrm{d}t} \\ \dfrac{\mathrm{d}\tilde{C}^m_i}{\mathrm{d}t} = D\left(\tilde{C}^{m,E}_i - \tilde{C}^{m,S}_i\right) + R^m_i(.) - \Phi^0_{i,m}(c) \end{cases} \tag{3.3.12a, b}$$

(Note that the usual case often implies that $R^m_i(.) = 0$.)

Comment In a dense medium, it can be admitted that, statistically, the number of micelles that form is equal to the number of micelles that disappear. The total number of micelles from a dispersing phase is therefore constant and the variational term vanishes in this case. However, the relation (3.3.4) may no longer be relevant in a dense medium.

3.3.3 Steady States in Simple Situations

It can be useful to deal directly with steady states that are a feature of the general, implicit equation for the mass balance because several very general forecasts can be made. Moreover, two small applications illustrate this at the end of this part of the thesis. This section shows how these steady states can be obtained.

For the requirements of his chapter, the general equation for the balances is written in a form that uses the sign for the different terms on the right of the equation.

$$\frac{\mathrm{d}\tilde{C}_i^c}{\mathrm{d}t} = -D\tilde{C}_i^c + \sigma_{qi}\left|q_i^c X^c\right| + \sigma_{\Phi i}\left|\Phi_{i,m}^c(c)\right| + \tilde{C}_i^c \frac{\mathrm{d}\ln N_T^c}{\mathrm{d}t} \tag{3.3.13}$$

where σ_{wz} is the denotation (−1, 0, 1) for the quantity wz. The quantities between bars, $|\ldots|$ indicate that these are considered in terms of absolute values.

3.3.4 Partial, Specific Rates

By examining Eq. (3.3.13), bearing in mind that the denotations (σ) can each take three values (−1, 0, 1), it is seen that there are, in theory, a possible nine forms for (3.3.13). In the case of the biphasic chemostat used for the working conditions defined above, only five cases have a nontrivial physical sense in the steady state. Table 3.2 takes up these five relevant cases and summarizes the most significant situations that they correspond to.

Let us consider briefly the form of the steady states corresponding to these five cases.

Table 3.2 The five relevant forms of the general mass balance

Case	$\sigma_{\Phi i}$	σ_{qi}	Physiological situation
A	0	+1	Intracellular synthesis without exchange with the external medium. E.g., ADN, ARN, etc.
B	+1	0	Passive stock (without change). E.g., ions
C	+1	+1	Collection (*uptake*) and simultaneous synthesis. E.g., amino acids
D	+1	−1	Collection (*uptake*) and simultaneous consumption. E.g., glucose…
E	−1	+1	Synthesis and simple excretion. E.g., exoenzymes

3.3.4.1 Case A. Strict Intracellular Synthesis

Here, it is a question of metabolites that are synthesized in the cell and which are not subject to any exchange with the other phases from which they are absent. This very important case concerns mainly the macromolecular cell compounds such as nucleic acids and the majority of proteins. In this case, the system from the equation describing the biphasic chemostat is,

$$\begin{cases} \dfrac{d\tilde{C}_i^c}{dt} = -D\tilde{C}_i^c + q_i^c X^c + \tilde{C}_i^c \dfrac{d\ln N_T^c}{dt} \\ \dfrac{d\tilde{C}_i^m}{dt} = 0 \end{cases} \tag{3.3.14a, b}$$

Comments In certain examples (as with system (3.3.14a, b) above), the quantities σ are replaced by their numeric values. So, (3.3.14a) is equivalent to (3.3.13) with $\sigma_{\Phi i} = 0$ and $\sigma_{qi} = +1$.

For the system (3.3.14b), in steady state, the number of cells is constant and the one condition of the steady state is,

$$q_i^c X^c = D\tilde{C}_i^c \tag{3.3.15}$$

This result shows that partial, internal flux from the synthesis of a strictly intracellular metabolite, is equal to the product of the dilution rate and the intracellular concentration of this metabolite. This explains why, in a general manner, the composition in cell macrocomponents must adjust with D if a steady state is to be maintained (Thierie 1997).

However, (3.3.15) can be put in the form:

$$q_i^c = \alpha_i^c D \tag{3.3.16}$$

This relation is particularly interesting because it expresses that the partial, specific synthesis rate of a metabolite (or group of metabolites) is simply proportional to the dilution rate, the coefficient for proportion, α_i^c, being no other than the mass fraction of the metabolite under consideration (the ratio of the mass of this metabolite in the cell phase, to the total cell mass). So, the evaluation of the net, specific rate of a strictly intracellular metabolite can be easily deduced from the content of this metabolite within the biomass at a given dilution rate.

The concept of mass fraction of a constituent in a micelle will be used several times. By definition,

$$\alpha_Z^p = \frac{m_Z^p}{m^p} \tag{3.3.17}$$

(For more details, see Appendix A.2).

3.3.4.2 Case B. Passive Storage

Here, it is a question about a case where a compound is simply extracted from the matrix medium and does not undergo any change (neither physical, chemical, nor biological) both inside the cell and in the matrix phase. If the collected compound is part of the influent, the system corresponding to this situation is hence,

$$\begin{cases} \dfrac{d\tilde{C}_i^c}{dt} = -D\tilde{C}_i^c + \Phi_{i,m}^{\circ}(c) + \tilde{C}_i^c \dfrac{d \ln N_T^c}{dt} \\ \dfrac{d\tilde{C}_i^m}{dt} = D\left(\tilde{C}_i^{m,E} - \tilde{C}_i^m\right) - \Phi_{i,m}^{\circ}(c) \end{cases} \tag{3.3.18a, b}$$

where $\tilde{C}_i^{m,E}$ is the pseudo-homogeneous concentration i at the entry to the reactor. The steady states are easily obtained. From (3.3.18a) it is calculated that

$$\Phi_{i,m}^{\circ}(c) = D\tilde{C}_i^c$$

and from (3.3.18b) that

$$\Phi_{i,m}^{\circ}(c) = D\left(\tilde{C}_i^{m,E} - \tilde{C}_i^m\right).$$

By combining these two relationships, it is seen that

$$\tilde{C}_i^c = \tilde{C}_i^{m,E} - \tilde{C}_i^m \tag{3.3.19}$$

This expression is nothing other than a simple expression of the conservation mass i in the system and nothing can be deduced as regards the relationship, the ratio of reconcentration of the compound in the cell. It presents, however, some interest since it predicts that the intracellular concentration of a compound only has passive storage that is independent of the dilution rate. This is a simple way to check whether a compound (toxic, for example) that penetrates a cell, undergoes or does not undergo, any sort of change. Any variation of intracellular concentration in accordance with D leads to suspicion of a transformation mechanism (even if it is a question of basic physisorption).

3.3.4.3 Case C. Collection and Simultaneous Synthesis

In this case, a compound (or group of compounds) can be both synthesized inside the cell and collected in the matrix phase (where it undergoes no change).

The corresponding system is written, for a component of the influent:

$$\begin{cases} \dfrac{d\tilde{C}_i^c}{dt} = -D\tilde{C}_i^c + \Phi_{i,m}^{\circ}(c) + q_i^c X^c + \tilde{C}_i^c \dfrac{d \ln N_T^c}{dt} \\ \dfrac{d\tilde{C}_i^m}{dt} = D\left(\tilde{C}_i^{m,E} - \tilde{C}_i^m\right) - \Phi_{i,m}^{\circ}(c) \end{cases} \tag{3.3.20a, b}$$

By eliminating $\Phi^{\circ}_{i,m}(c)$ from the steady solutions of these two equations, it is seen that

$$q_i^c = D\alpha_i^c - \frac{D\left(\tilde{C}_i^{m,E} - \tilde{C}_i^m\right)}{X^c}$$

and using a yield coefficient,

$$Y_{X^c,i} = \frac{X^c}{\tilde{C}_i^{m,E} - \tilde{C}_i^m} > 0$$

the partial, specific synthesis rate takes the form,

$$q_i^c = D\left(\alpha_i^c - \frac{1}{Y_{X^c,i}}\right) \geq 0 \qquad (3.3.21)$$

(Negative values for q_i^c arise from another case.)

By comparing this relationship with (3.3.16), it is noted that where α_i^c is constant, the net partial, specific rate of synthesis of a compound that can be collected in the matrix phase is less than it would be if the compound were synthesized completely in the cell; in fact,

$$D\left(\alpha_i^c - \frac{1}{Y_{X^c,i}}\right) < D\alpha_i^c$$

In the case of intracellular synthesis, concomitant with uptake of the same compound in the matrix phase, it is to be expected that the compound taken up has an effect comparable to an inhibition or suppression on its own synthesis rate. This result is effectively observed in the case of biosynthesis of amino acids (AA). When there is an upshift in the growing of a culture on a minimal medium (without AA) toward a medium enriched with AA, a rapid incorporation of AA by the cell is observed, followed by the repression of enzymes involved in the biosynthesis of these (Ingraham et al. 1983). In the minimal medium, the formation of a pool of AA is strictly intracellular and is governed by (3.3.16); in the enriched case, on the contrary, exchange with the matrix phase takes place, leading to enzymatic repression that diminishes the partial, specific synthesis rate in accordance with (3.3.2). (Note, however, that the enzymatic repression observed in the biosynthesis of AA is not the only mechanism possible to satisfy (3.3.2), but examination of other cases has not been undertaken.)

3.3.4.4 Case D. Simultaneous Uptake and Consumption

This situation involves the incorporation and simultaneous change of the compounds in the cell. This case is obviously extremely important since it includes all the nutrients. The corresponding system is

$$\begin{cases} \frac{\mathrm{d}\tilde{C}_i^c}{\mathrm{d}t} = -D\tilde{C}_i^c + \Phi_{i,m}^{\circ}(c) - q_i^c X^c + \tilde{C}_i^c \frac{\mathrm{d}\ln N_T^c}{\mathrm{d}t} \\ \frac{\mathrm{d}\tilde{C}_i^m}{\mathrm{d}t} = D(\tilde{C}_i^{m,E} - \tilde{C}_i^m) - \Phi_{i,m}^{\circ}(c) \end{cases} \qquad (3.3.22\text{a, b})$$

In the same way as in the previous case, the following relationship at steady state is obtained:

$$q_i^c = D\left(\frac{1}{Y_{X^c,i}} - \alpha_i^c\right) \geq 0 \qquad (3.3.23)$$

The specific rate of degradation of the metabolite is thus diminished by the presence of the free metabolite in the cell. This situation can be unfavorable if the metabolite concerned is part of the compartment linked to the production of energy (*fuelling*) of the cell. In this case, it can be expected that the possibility of obtaining maximum energy in minimum time (at least potentially, in form of precursors) is desirable in some cases. With the phenomenon of constant transfer, optimization of the substrate rate of use, for example, is

$$\max(q_i^c) = \frac{D}{Y_{X^c,i}}$$

which implies that $\alpha_i^c = 0$.

We can assume that phenomena of transfer and use are of the same order of magnitude and are very rapid in comparison with hydraulic phenomena and cell multiplication. This situation is called the quasistationary hypothesis (QSH) and is expressed theoretically by,

$$\Phi_{i,m}^{\circ}(c) \equiv q_i^c X^c \qquad (3.3.24)$$

In these conditions, Eq. (3.3.22a) is reduced to

$$\frac{\mathrm{d}\tilde{C}_i^c}{\mathrm{d}t} = \tilde{C}_i^c\left(\frac{\mathrm{d}\ln N_T^c}{\mathrm{d}t} - D\right) \qquad (3.3.25)$$

of which the steady states are

$$\frac{\mathrm{d}\ln N_T^c}{\mathrm{d}t} = D \qquad (3.3.26\text{a})$$

and

$$\tilde{C}_i^c = 0 \qquad (3.3.26\text{b})$$

Solution (3.3.26a) is very interesting, but does not apply to the biphasic committee that has only one species of cell. Therefore (3.3.26b) remains, and this obviously implies that $\alpha_i^c \equiv 0$. It can be shown that the QSH has an important consequence regarding the expression of growth rate for the whole of the cell phase (specific rate at the level of biomass ("µ")): it in fact implies that the endogenous flux of this compound is zero and that its mass contribution to the rate of the whole is nil. Obviously, this in no way means that the role of this metabolite in cell growth is negligible, but only that in this form i does not contribute significantly the production of biomass. On the contrary, its energy properties, or those of its derived forms, can have considerable importance.

It is interesting to note that the relationship (3.3.23) is identical, or nearly identical to the relationship (3.3.21) and that

$$q_i^c(\text{case}\, C) = -q_i^c(\text{case}\, D)$$

The specific rate of synthesis can thus be theoretically defined on an axis interval $]-\infty, +\infty[$, the change in denotation obviously corresponding to a change of physiological situation (a state transition phenomenon which will be looked at later).

3.3.4.5 Case E. Synthesis and Excretion

The compound synthesized in the intracellular medium can be excreted in the matrix phase and may or may not undergo change. When the compound excreted does not undergo change it is called simple excretion. In the case treated here, it is assumed that the product excreted is absent from the reactor feed (influent), so,

$$\begin{cases} \dfrac{d\tilde{C}_i^c}{dt} = -D\tilde{C}_i^c - \Phi_{i,c}^{\circ}(m) + q_i^c X^c + \tilde{C}_i^c \dfrac{d \ln N_T^c}{dt} \\ \dfrac{d\tilde{C}_i^m}{dt} = -D\tilde{C}_i^m + \Phi_{i,c}^{\circ}(m) - R_i^m(.) \end{cases} \tag{3.3.27a, b}$$

where $R_i^m(.)$ is a term for the consumption of compound in the matrix phase.

The partial, specific rate of synthesis in the steady state is easily obtained,

$$q_i^c = D\left(\alpha_i^c + \frac{\tilde{C}_i^m}{X^c}\right) + \frac{R_i^m(.)}{X^c} \tag{3.3.28}$$

First of all, let us look at a simple excretion ($R_i^m \equiv 0$). By comparing with (3.3.16), for the same mass fraction, the rate of synthesis i is bigger in the case of excreted product. This result makes sense because the synthesis must now be able not only to provide for the intracellular concentration, but also for that of the exterior medium. When $R_i^m > 0$, the rate of synthesis must still increase, for α_i^c and $\tilde{C}_i^m$ constants, to compensate the disappearance of the compound from the medium.

The excretion phenomena, by increasing the rate of synthesis, can have a large energy cost. This cost can be compensated for if it allows supplementary nutrients to be brought in (as in the case if certain exoenzymes) or be justified if the excretion is used to eliminate toxic or osmotically dangerous subproducts. However, from the biotechnological point of view, the most interesting phenomenon is without doubt, the one regarding the possibility of interaction between intracellular production and the reaction in the center of the same compound. For example, a phenomena of excretion that might depend on a gradient of concentration between the interior and exterior of the cell would be affected by a decrease in the extracellular concentration. Not only would the production be improved by it, but the constraint imposed on the specific synthesis rate could induce changes in physiological state affecting the whole cellular phase. From this perspective, the action on the excreted products concentration in the medium could constitute an extra cellular regulating parameter at the system level. At present the phenomena of excretion are thought to be insufficiently studied from this point of view and perhaps in general too.

General comment The five cases studied above represent in a way, the archetypes of five situations (out of nine) that are relevant for the biphasic chemostat. Synthesis of results appears in Tables 3.2 and 3.3.

The most complex systems, notably those that are made up of a large number of phases, present a much greater number of situations. However, even in cases which are under consideration, bear in mind that each partial, specific rate or each exchange flux is (or could be) a net quantity, that is to say, the sum of several terms. By way of example, if the synthesis rate of the protein family of compounds q^c_{PSS} (*Protein Synthesis System*), is taken into consideration, it is obvious that this term, in the case of a bacterium, is the sum of a thousand terms each corresponding to the

Table 3.3 Mass balances corresponding to the five relevant cases

Case	$\frac{\mathrm{d}\tilde{c}^c_i}{\mathrm{d}t} =$	$\frac{\mathrm{d}\tilde{c}^m_i}{\mathrm{d}t} =$
A	$-D\tilde{C}^c_i + q^c_i X^c + \tilde{C}^c_i(\mathrm{d}\ln N^c_T/\mathrm{d}t)$	0
B	$-D\tilde{C}^c_i + \Phi^\circ_{i,m}(c) + \tilde{C}^c_i(\mathrm{d}\ln N^c_T/\mathrm{d}t)$	$D(\tilde{C}^{m,E}_i - \tilde{C}^m_i) - \Phi^\circ_{i,m}(c)$
C	$-D\tilde{C}^c_i + \Phi^\circ_{i,m}(c) + q^c_i X^c + \tilde{C}^c_i(\mathrm{d}\ln N^c_T/\mathrm{d}t)$	$D(\tilde{C}^{m,E}_i - \tilde{C}^m_i) - \Phi^\circ_{i,m}(c)$
D	$-D\tilde{C}^c_i + \Phi^\circ_{i,m}(c) - q^c_i X^c + \tilde{C}^c_i(\mathrm{d}\ln N^c_T/\mathrm{d}t)$	$D(\tilde{C}^{m,E}_i - \tilde{C}^m_i) - \Phi^\circ_{i,m}(c)$
E	$-D\tilde{C}^c_i - \Phi^\circ_{i,c}(m) + q^c_i X^c + \tilde{C}^c_i(\mathrm{d}\ln N^c_T/\mathrm{d}t)$	$-D\tilde{C}^m_i + \Phi^\circ_{i,c}(m) - R^m_i$

Table 3.4 Steady states (SS) corresponding to the kinetics of Table 3.3

Case	$\sigma_{\Phi i}$	σ_{qi}	SS	Physiological situation
A	0	+1	$q^c_i = D\alpha^c_i$	Intracellular synthesis alone
B	+1	0	$\tilde{C}^c_i = \tilde{C}^{m,E}_i - \tilde{C}^m_i$	Passive storage
C	+1	+1	$q^c_i = D\left(\alpha^c_i - \frac{1}{Y_{X^c,i}}\right) \geq 0$	Uptake and synthesis
D	+1	−1	$q^c_i = D\left(\frac{1}{Y_{X^c,i}} - \alpha^c_i\right) \geq 0$	Uptake and consumption
E	−1	+1	$q^c_i = D\left(\alpha^c_i + \frac{\tilde{C}^m_i}{X^c}\right) + \frac{R^m_i(.)}{X^c}$	Synthesis and excretion

synthesis of a specific protein, $q^c_{P_i}$. Each one of these terms itself corresponds to several dozen (perhaps hundreds) of terms, including the synthesis of amino acids, fuelling stages, etc. At a, probably, lower level of complexity, the same remark goes for the interphasic exchange flows. So, a reversible (or irreversible) inflow or outflow will be considered as an uptake if the net flow corresponds to the passage from the matrix toward the cell, but could appear as an excretion if the resulting flow corresponds to the passage from the cell toward the matrix. The result is that under various different constraints (different rates of dilution, for example) changes occur from one case to another.

To illustrate this proposition, different cases can be represented in the plane $\Pi(q^c_i, \alpha^c_i)$.

The straight line d corresponds to case A of strictly intracellular synthesis $d \equiv q^c_i = D\alpha^c_i$. It is a line passing through the origin and of slope D. Let us imagine an invariant point in the plane ($q^c_i, \alpha^c_i =$ ctes.). For dilution rate D1, synthesis is intracellular. By decreasing D (D3 < D1), the cell is to go into a situation of excretion (area above $q^c_i = \mathrm{D3}\alpha^c_i$). By increasing D (D2 > D1) the system should revert to a uptake/synthesis situation (that is to say, to maintain $q^c_i, \alpha^c_i =$ ctes., the medium will have to be supplemented with compound i and the cell made impermeable to this (cf. Fig. 3.5A (box)). It can easily be imagined that variation in

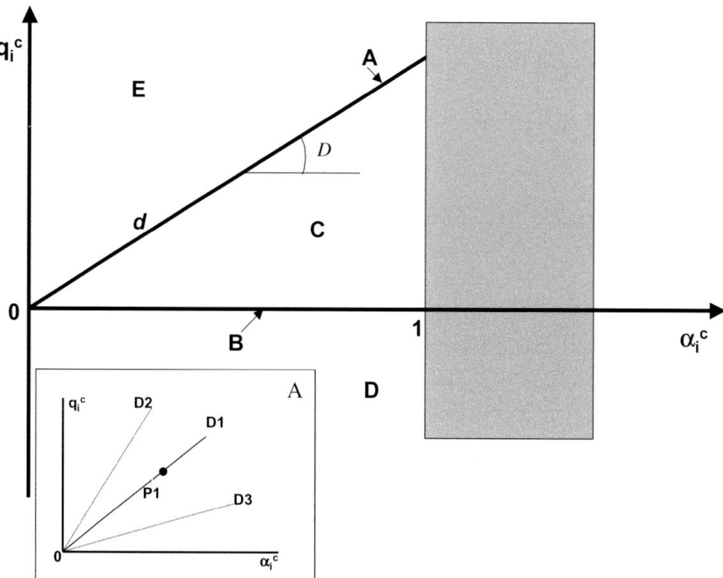

Fig. 3.5 The five physiological situations from Table 3.4 in the plane $\Pi(q^c_i, \alpha^c_i)$. The graph shows the different domains from area $\Pi(q^c_i, \alpha^c_i)$ determined by case A (*straight line d*) for a given dilution rate, D. Box A shows the position relative to an invariable point P1, for an increase or a decrease of D = D1. So for D3 < D1, P1 belongs to a domain of type E; for p D2 > D1, P1 belongs to a domain of type C

D will tend to lead to a variation q_i^c and/or α_i^c. In particular, for all strictly intracellular synthesis, it will be impossible to vary D while preserving these two constant values *at the same time.* In other terms, the variation of dilution rate defines a **trajectory** in the plane. Each point on this trajectory corresponds to a **mean physiological state** particular to the cells. The study of this trajectory, its continuity properties, its pathway, etc., could lead to a great deal of information on cell behavior of a system subject to hydraulic constraints.

3.3.5 The Water Problem

The basis of the PDS theory rests on the mass balances of compounds within each phase. In general, the relevant compounds studied are proteins, nucleic acids, sugars, etc. However, from the mass point of view, the main compound of the cell is water estimated at a minimum of 70 % of the bacterial cell mass, for example.

Water is not only a compound that easily crosses through the cell membrane in one direction or another (interphasic exchange flow), but it is also largely synthesized (produced) in intracellular medium.

By way of example, the stoichiometry of global synthesis of *Enterobacter* (*Aerobacter*) *aerogenes* growing on glycerol and ammonia (that will be considered later) is

$$C_3H_8O_3 + aO_2 + bNH_3 \rightarrow cCH_{1.83}O_{0.55}N_{0.25} + dCO_2 + eH_2O$$

With a = 2.157; b = 0.344; c = 1.375; d = 1.625; e = 3.258, coefficients calculated on the basis of RQ = 0.753 (Doran 1995; RQ = d/a is the respiratory quotient).

In terms of mass, this means that about 167 g of substrates provide 36 g of biomass, 72 g of CO_2, and 59 g of water, if there is only 22 % of biomass against 35 % of water (and 43 % of CO_2). The order of magnitude of the synthesis of biomass compared with that of water is nearly from single to double. Neither the exchange flux nor the specific synthesis rates of water can be considered as negligible in a mass balance relating to microorganisms. (The role of water is obviously also very important from a physiological point of view, independent of mass balances, but that is not the subject here.)

Having taken into account these observations, it is convenient to define carefully several of the terms used above. Biomass X^c is the **total** biomass, the sum of all the cell constituents, including water. The dry weight is not equal to X^c. Using M_w^c to denote the average mass of water in the cell and X'^c the dry weight and α_w^c the average water content (mass fraction) in the cell phase, taking into account that $X^c = M^c/V_T$, it is seen that

$$X'^c = \frac{M^c - M^c_w}{V_T} = X^c - \frac{M^c_w}{V_T}$$

$$X'^c = X^c\left(1 - \frac{M^c_w}{V_T X^c}\right) = X^c\left(1 - \frac{M^c_w}{M^c}\right)$$

and by definition,

$$X'^c = X^c(1 - \alpha^c_w) \tag{3.3.29}$$

just as

$$\alpha'^c_i = \frac{M^c_i}{M^c - M^c_w} = \frac{M^c_i}{M^c(1 - \alpha^c_w)}$$

and so,

$$\alpha'^c_i = \alpha^c_i \frac{1}{1 - \alpha^c_w} \tag{3.3.30}$$

where α'^c_i is the fraction of mass calculated relative to the dry weight (and not relative to the total biomass). It results from this that the partial, specific rates cannot be evaluated directly from the dry weights, at least without precise knowledge of the water, according to D. In retrospect, from the fact that $X'^c\alpha'^c_i = X^c\alpha^c_i$ (cf. (3.3.30) and (3.3.29)), the partial intracellular transformations fluxes can be calculated accurately since evidently,

$$q^c_i X^c = D\alpha^c_i X^c = D\alpha'^c_i X'^c \tag{3.3.31}$$

3.3.6 Examples

Using two examples, the exploitation of the results given above will be illustrated briefly. The accent is put more on the use of PDS theory than on in-depth interpretations of the microbiological phenomenon. So, by example, the methodology of the experiment can be questioned without hindering the application of theory. In spite of this, and although the examples are used to illustrate a methodology, the results are rapidly interpreted to show the improvement that our approach can possibly bring as compared to the classical approach.

3.3.6.1 Example a. The Partial, Specific Synthesis Rate of Cytochromes in *Pseudomonasfluorescens*

The experimental results used here are the fruit of a work by Rosenberger and Kogut (1958). *Pseudomonas fluorescens* (strain KB1) is cultivated in chemostat in a medium consisting of succinate, ammonium, and salts. The quantity of cytochromes contained in the culture is obtained using a measurement of the difference in optical density (OD) of 418 nm, maximum absorption of the cytochromes, and of the optical density of 500 nm that is supposed to represent the adsorption of the other constituents and diffused light. According to the methodology, the measurement represents the relative quantity of cytochromes in suspension. In fact, it was not possible to construct a standard curve cytochromes from strain KB1. For this reason, the results are expressed in OD. On the other hand, it was possible to obtain a linear relationship between OD (415) and OD (500) and the variation in concentration of cytochrome *c* (from horse heart) that was added. This standard curve does not go through the origin (Rosenberger and Kogut 1958, Fig. 3). Two experimental curves for DO (418)–DO (500) in the steady state, according to the dilution rate have been observed, one under conditions where the succinate is the limiting factor and the other where the limiting factor is the air provided to the culture. The results are represented in Fig. 3.6.

The following hypotheses have been made:

(i) the methodology used is nondestructive and achieves a measure of the complete and intact cells. From this fact, it is supposed reasonable to consider that the measurement effected is linear in relationship to the fraction of mass, $\alpha^{c}_{\mathrm{CYT}}$ from strain KB1;

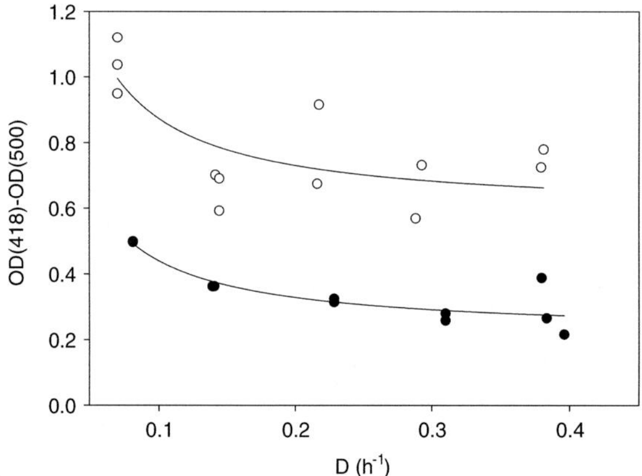

Fig. 3.6 Difference between OD at 418 and 500 nm according to dilution rate *D*. The difference between the two optical densities is expected to represent the relative cytochrome concentration relative to cytochrome in suspension. (The *black circles* indicate the concentrations in the culture limited by succinate, EXP A; the *white circles*, those estimated in the culture limited by air, EXP B)

Table 3.5 Results of the parametric estimation of (3.3.32)

n° exp.	Condition of culture	aK	b	r
A	Succinate limited	0.022 ± 0.004	0.218 ± 0.025	0.89
B	Air limited	0.028 ± 0.008	0.589 ± 0.066	0.75

(ii) the linear relationship between optic density and KB1 cytochrome content is of the same type as that obtained experimentally with cytochrome c (without this hypothesis, the methodology would not make sense). The straight line does not go through the origin.

(iii) the cytochromes measured are not excreted in the matrix phase.

In these conditions, the synthesis is strictly intracellular and must satisfy the relationship (3.3.16). Let

$$q^c_{\mathrm{CYT}} = D\alpha^c_{\mathrm{CYT}}$$

where q^c_{CYT} is the partial, specific rate of the family of cytochromes in the dose.

If this rate of synthesis is constant ($q^c_{\mathrm{CYT}} = \mathrm{cte} = K$), a relationship between α^c_{CYT} and D of the type $\alpha^c_{\mathrm{CYT}} = K/D$ should be observed. According to the hypotheses (i) and (ii), it is seen that

$$\mathrm{OD}(418) - \mathrm{OD}(500) = \frac{\mathrm{aK}}{\mathrm{D}} + \mathrm{b} \tag{3.3.32}$$

where aK and b are constants, characteristic of the system. The results of the fittings using least squares that were done on the two experimental series are shown again in Table 3.5 (continuous curves in Fig. 3.6).

The fittings obtained are realistic and perfectly compatible with the hypothesis that, for the two conditions of culture, the partial, specific rate of synthesis of the cytochromes from *Pseudomonas* KB1 is constant at all dilution rates. The data are nevertheless insufficient to determine their value. The coefficients $aK(A)$ and $aK(B)$ may not be very different, however, $b(B) \approx 3b(A)$ which shows that the culture conditions influence the specific production rate of cytochromes.

In the absence of further data (notably on the nature of the cytochromes under conditions A or B and variation of the OD(500)), it can scarcely be concluded that there is a high probability that the partial synthesis rate is constant. By making the (risky) hypothesis that the slopes of the standard curves in conditions A and B are the same, it is obtained that $q^c_{\mathrm{CYT}}(B) \approx 1.3 \times q^c_{\mathrm{CYT}}(A)$, which would confirm conclusions that there is an increase in the biosynthesis in the medium with limited air. This increase can then be estimated to be relative to the order of 30 %.

3.3.6.2 Example b. Synthesis of Proteins and RNAs from *Aerobacter aerogenes*

Variation in the composition of *Enterobacter* (or *Aerobacter*) *aerogenes* (NCTC418) cultivated in chemostat in the steady state at different rates of dilution in a medium with limited glycerol (C-limited) was compared to the same culture with limited magnesium (Mg-limited) (Tempest et al. 1965). On the basis of the results there is discussion (notably) on the possibility of regulation of synthesis of RNAs by magnesium and the efficacy of ribosomes in the synthesis of proteins in the two culture media. Comparison of the mass of RNA produced, expressed in homogeneous concentration, shows a very spectacular difference between the two media (Tempest et al. 1965) and leads to the conclusion that the synthesis of RNAs is effectively regulated by magnesium. It is suggested that there is a variation in the activity of the ribosomes.

These results were looked at again in order to examine them from the perspective of PDS. In order to have available a complete series of data for the two media, a polynomial fitting was done using the original, deficient results. The "leveling" of the data is very satisfying and gives residues that uniformly distributed over the interval of the rates of dilution studied experimentally. (The quantity of data concerning these fittings is quite important and does not bring in supplementary information of that to be illustrated. So as not to overload the text unnecessarily, these adjustments are not shown here.) The values for biomass being given as dry weight, the estimation of partial, specific rates of synthesis was not possible, even approximately, because the osmotic conditions of the two media are too different to make reasonable hypotheses on the water content of the cells. The principal hypothesis is that all compounds considered, are strictly intracellular productions. Therefore, the relationship (3.3.31) is used to calculate the partial change fluxes in a stringent way ($q_i^c X^c = D\alpha_i^{\prime c} X^{\prime c}$). The results are shown in Fig. 3.7 (C-limited) and in Fig. 3.8 (Mg-limited). Quantitatively, it is clear that absolute, general productivity

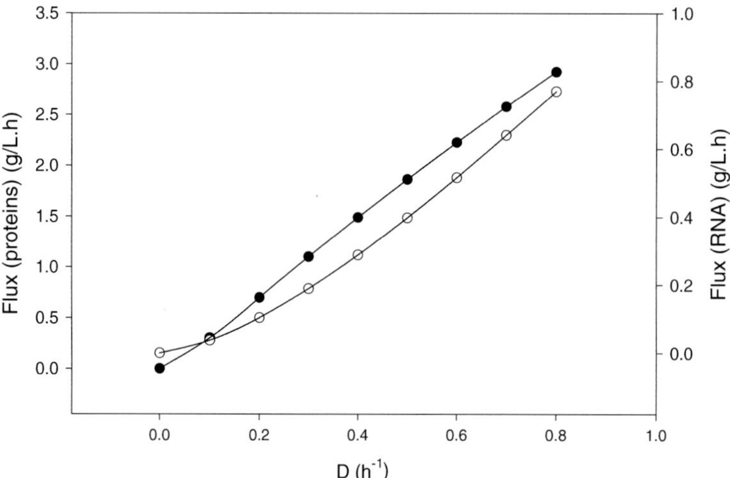

Fig. 3.7 Fluxes of proteins and RNAs in the medium with limited carbon. The flow of proteins was calculated according to the relationship $q_i^c X^c = D\alpha_i^{\prime c} X^{\prime c}$ using the interpolated, experimental values (by adjustment; solid lines). The experimental values are represented by dots (*black circle*) proteins, (*white circle*) RNA

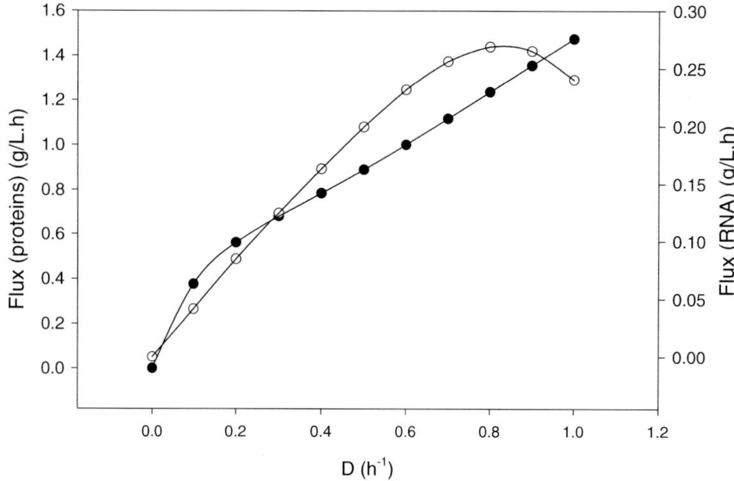

Fig. 3.8 Flux of proteins and RNAs in a medium with limited magnesium. The flow of proteins we calculated according to the relationship $q_i^c X^c = D\alpha_i'^c X'^c$ using the interpolated experimental values (by adjustment; *solid lines*). The experimental values are represented by dots (*black circle*) proteins, (*white circle*) RNAs

is less on the medium with limited magnesium, where the fluxes are more or less half that of those of the medium with limited glycerol.

Qualitatively, however, the differences are less marked, except for nondosed compounds ($q_{\text{OTH}}^c X^c$) that are obtained by the difference ($\alpha_{\text{OTH}}'^c = 1 - \sum_{i \neq \text{OTH}} \alpha_i'^c$). (The relatively higher proportion of these in the medium with limited magnesium suggests perhaps membranous compounds.) The most marked differences were in the difference between the curves of $q_{\text{PSS}}^c X^c$ and $q_{\text{ARN}}^c X^c$ that seem to be contrary signs in most of the domain of D. (Contrary to $q_{\text{ADN}}^c X^c$ and to $q_{\text{CAR}}^c X^c$ which preserve the same profile.)

On the subject of the discussion mentioned above, it seemed interesting to look at the relationship between the partial rates of changes in RNAs and proteins (PSS—*protein synthesis system*). This relationship gives, in fact, the relative proportion of the partial, specific synthesis rate,

$$\frac{q_{\text{ARN}}^c X^c}{q_{\text{PSS}}^c X^c} = \frac{q_{\text{ARN}}}{q_{\text{PSS}}^c}$$

Figure 3.9 shows the variation of this relationship as a function of D and shows that:

1. this relationship is an increasing function of the dilution rate;
2. the values obtained are not significantly different in the two media. (The deviations compensate for each other and are of the order of a percentage of an average value, a difference that is perfectly compatible with the dispersion of experimental results.)

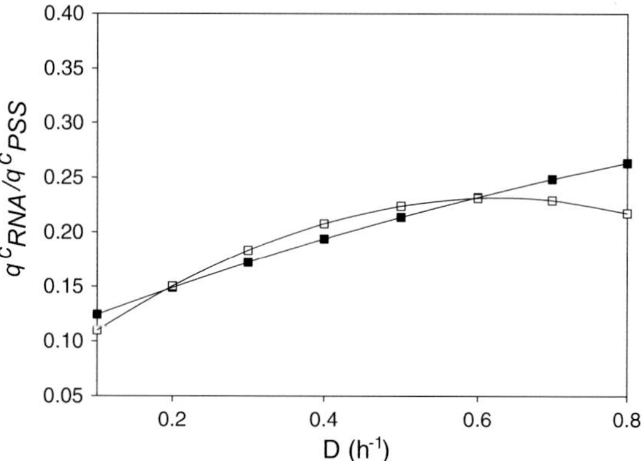

Fig. 3.9 Relationship of the partial specific rates of RNAs and proteins in the two media (*black square* limited carbon medium; *white square* limited magnesium medium)

Given the results, it can be admitted that there is a variation just about linear of the ratio with the dilution rate. The adjustment of all points by the least squares method gives 0.18 D + 0.12 (r = 0.94). The slope shows an increase of 0.18 (g_{RNA}/g_{PROT}) • h, a value very close to that obtained by Cole et al. (1987) who give 0.2 (g/g) • h (r = 0.9) for *Escherichia coli.*

These results suggest that, independent of a possible regulation of the synthesis of the RNAs by Mg^{2+} and/or a difference in the efficacy of the ribosomes, the net global co-regulation of total RNAs and proteins synthesis is the same in the two media. This conclusion is valid if the RNAs and the proteins are, in both media, strictly intracellular compounds, such as is usually the case. Nevertheless, this had not been proven experimentally in the example studied here.

The two examples above clearly show the efficiency of the PDS representation of various results. However, the cases treated teach also that in order to make effective use of the general results, it is useful to provide an experimental protocol that is a little different from that usually used. In particular, the expression of result only in dry weight limits severely the bearing on interpretations, particularly in terms of absolute, partial, specific rate. The precise determination of the water content in the microorganisms is therefore an important factor in the PDS theory. At present, this determination is often delicate and if not really a part of routine measurements. So, the development of a simple and effective method to determine the fraction of water in the biomass seems desirable.

3.4 The Concept of Maintenance

Preliminary Notes

When the general mass balances such as (2.4.26) or (3.3.7) are used, this is called the application of implicit or semi-explicit forms' balances. In particular, the biphasic chemostat is described in Sect. 3.3 in a semi-explicit form that only explains the inlet and outlet, hydraulic terms of the bioreactor. As these terms are very simple in the case of the chemostat, it is moreover considered that the description of the mass is "implicit," simply when the specific rates are not explained. This is the case in this section that illustrates the efficacy of the description for polyphasic dispersed systems, even when they are used in an implicit form.

3.4.1 Introduction

By paraphrasing Kuenen's (1979) definition, the term "maintenance energy is used for all forms of energy necessary for growth and used in such a way that this energy is not available for biosynthesis." Although already presented in 1942 by Monod (1942), the concept of maintenance was formally introduced by Pirt (1965). In this first model by Pirt, the maintenance energy was independent of the growth rate. This model was tested in numerous cases (Pirt 1982). However, it appears that numerous systems showed maintenance energy that varied with growth rate (first by Neijssel and Tempest 1976). Several attempts to improve Pirt's initial model were made (Neijssel and Tempest 1976; Pirt 1982; Tsai and Lee 1990). What distinguishes the two ways of achieving maintenance (constant or variable maintenance) seems to concern mainly the conditions of growth and, in particular, whether the culture medium is limited or not, by the principal, carbonated substrate.

Diagrammatically, the maintenance energy would be constant in systems where provision of carbon (and/or energy) is sufficient, and variable in systems that have enough carbon (and/or energy). At the present time, it can be considered that the proposed models make it possible to take into account, experimental observations, in a phenomenal way. The difficulty rests in the physiological interpretation of the models. A fundamental criticism of the models described above, is aimed at raising all precise, biological significance to quantities measured using these models (Tempest and Neijssel 1984). Such a criticism reverts to displaying the ad hoc character of the formulae obtained so it would be more a question of curve fitting than of physiologically relevant models. In his article of 1982, Pirt himself stated that "the explanation of maintenance energy requirements remains largely a physiological microbial challenge." Today, it appears that this statement is more than ever a current one because these models are very largely used in the modeling of biotechnological procedures. So it is important, to go beyond a simple measurement of the parameter of maintenance and to *understand* the phenomena being dealt with, in order to optimize and control procedures.

In this section, attempts are made to construct an approach that is very different from those that precede. However, the scope is not beyond the case of growth limited by the main carbon course. There are several reasons for this. In the models described above, as in those of Liu and Chen (1997), the specific rate of consumption of substrate is considered as the only quantity that should be taken into account to evaluate the maintenance terms. This hypothesis seems fully justified in the case of insufficient carbon for which the correlation between growth and the limiting substrate seems to go without saying. The hypothesis seems more problematic when the limiting substrate is not carbonated substrate. In these cases, it would seem reasonable to take into account the consumption rate of limiting substrate too. One other reason stems from the fact that, with the problem called "maintenance"' "the crux of the problem rests in the apparent capacity of microorganisms to totally or partially dissociate catabolism from anabolism, when circumstances dictate" (Tempest and Neijssel 1984). This problem of division is fundamental in this field. Now, it appears that in cases where there is sufficient carbon substrate, numerous, metabolic by-products are secreted, but not when the carbonated substrate is limiting (Neijssel and Tempest 1975). So, it appears that different metabolic pathways are brought into use, according to the culture conditions. Such a situation can only obscure the problem of division by leaving incomparable, the growth situations that are relevant of the different, physiological situations. Over this, we perfectly agree with the conclusions of Tempest and Neijssel (1984) who assert that noncarbon-limited cultures are intrinsically different from carbon-limited cultures. So the choice was made to limit the approach to carbon-limited cultures not including secretion of metabolic by-products. This situation seems to correspond to the simplest physiological case that could be considered. Now, it remains to determine under which conditions such a simple physiology is likely to give rise to complex phenomena that are usually explained in terms of maintenance.

From a biotechnological point of view, the maintenance energy in microbial cultures can be, like the language of Aesop, both the best and the worst thing. For a biomass production exclusively, as in the case of baker's yeast or single-cell proteins, the optimization of the process aims to reduce energy waste as far as possible. In an environmental bioremediation process, such as biological wastewater treatment, the ideal procedure would be the one that wasted most energy, the produced biomass being a nondesirable solid waste. If it is a question of producing useful overflow by-products from the metabolism (alcohols, solvents, antibiotics), maintenance energy is advantageous (Russel and Cook 1995) but cell growth can also intervene favorably (Doran 1995) and "the interdependence of the product that comes from formation and microbial growth is a determining factor in the optimization of productivity yield" (Hollander 1993). Optimization then passes through a compromise between waste energy and production of biomass.

The profits of the fermentation industry, for example, are mainly dependent on the efficiency gained from the process of fermentation itself and in spite of considerable efforts made in the most ancient of technologies, the strains of yeast are still very far from being optimum (Walker 1998). Ethanol production process, for example, is extremely sensitive in terms of profit, to the improvement and use of the growth substrate that makes up the biggest part of production costs. Glazer and Nikaido (1995) confirm that an increase in yield of ethanol from 90 à 92 % can lead to a fall in production costs of 1 % or more.

It is clear from what has just been said, that waste of energy plays a very important role in bioengineering, as much from the point of view of procedures efficiency as from that of their profitability. The example of ethanol production shows well that in certain cases, an improvement of just a few percent in the effectiveness of a procedure that is already very efficient, can have considerable consequences from an economic point of view. This makes it to be thought that in terms of maintenance energy, whether it is increased or decreased, there are no small advantages. From a practical point of view, in production, for example, the description of a bioreactor on several levels is interesting. On one hand, global quantities (biomass, residual substrate, etc.) can be followed in the usual manner. However, on the other hand, it is possible to take into account in the same representation of local kinetics that occur at cell level. The recent development and rapid progress in particular probes make it more or less possible to refine monitoring of these processes and also to increase their performances by improving strains, this improvement being linked to a better understanding of cell mechanisms and above all to the possibility of measuring them in real time (Kwong and Rao 1994; Blake-Coleman 1993).

3.4.2 Method

The general method for obtaining the mass balance is described in Sect. 3.2. Only the concepts that apply to the biphasic chemostat, defined in Sect. 3.3, are used here.

It is not useless to reframe the main, general results obtained above and to insist on certain concepts that will be useful for the following development of the subject matter.

The law of evolution of the carbon-limited substrate mass balance, S, in the cell phase is given by,

$$\frac{\mathrm{d}\tilde{C}_S^c}{\mathrm{d}t} = -D\tilde{C}_S^c + \Phi_{\text{obs}}^0 - q_S^c X^c + \tilde{C}_S^c \frac{\mathrm{d}\ln N_T^c}{\mathrm{d}t} \tag{3.4.1}$$

$\tilde{C}_S^c$ is pseudo-homogeneous concentration of the substrate associated with the cell phase; D is the dilution rate (Q/V_T); q_S^c is the net specific rate of disappearance of the substrate in the cell phase. It represents the sum of the processes that consume the substrate in the cell phase (it includes the phenomena of transport and metabolism). In a general manner, q_S^c can be put in the form,

$$q_S^c = \sum_i q_{S,i}^c \tag{3.4.2}$$

where

i represents a subgroup of given processes;

X^c is the total biomass;

N_T^c is the number of cells that form the biomass.

Note that the unusual formalism for interphasic exchange flow is Φ^0_{obs}, which simply corresponds to the usual definition (3.3.8); the reason for this change in denotation will appear later.

In the matrix phase, the law of evolution of the substrate is given by,

$$\frac{\mathrm{d}\tilde{C}_S^m}{\mathrm{d}t} = D\left(\tilde{C}_S^{m,E} - \tilde{C}_S^m\right) - \Phi^0_{\text{obs}} \tag{3.4.3}$$

$\tilde{C}_S^{m,E}$ is the concentration of the substrate in the matrix phase at the inlet to the reactor and $\tilde{C}_S^m$ is the concentration in the core of the reactor;

Φ^0_{obs} is the net substrate exchange flux per unit of volume, between the matrix phase and the cell phase.

All values given above are positively defined.

In the steady state, the variational terms are canceled and from (3.4.1) one obtains,

$$q_S^c X^c = \Phi^0_{\text{obs}} - D\tilde{C}_S^c \tag{3.4.4}$$

Equation (3.4.3) gives the following stationary state:

$$\Phi^0_{\text{obs}} = D\left(\tilde{C}_S^{m,E} - \tilde{C}_S^m\right) \tag{3.4.5}$$

The net exchange flow, Φ^0_{obs}, is in reality simply the *observable* quantity of substrate that leaves the matrix phase to enter into the cell phase.

3.4.3 *Results*

3.4.3.1 Principle of the Method

The description of the principle of the method is perhaps more important than a rigorous demonstration. We will explain the purpose of our approach before seeking complete and precise analytical expressions.

The yield coefficient of the biomass in relation to the substrate is by definition,

$$Y_{X^c,S} = \frac{X^c}{\tilde{C}_S^{m,E} - \tilde{C}_S^m} \tag{3.4.6}$$

which measures the quantity of biomass formed per unit of substrate consumed. The variation of this coefficient can therefore be made either by variation of the substrate consumed with constant biomass, or by variation in the biomass with constant substrate (or by a combination of the two). We consider the case where the constraint is that the substrate consumed is constant, that is,

$$\tilde{C}_S^{m,E} - \tilde{C}_S^{m} = \text{cte.} \tag{3.4.7}$$

Let us consider the mass balance (3.4.4) in the following form:

$$\Phi_{\text{obs}}^{0} = q_S^c X^c + D\tilde{C}_S^c \tag{3.4.8}$$

If the concentration of free substrate in the cell phase is very small in comparison with the intracellular flux, (3.4.8) takes the simplified form,

$$\Phi_{\text{obs}}^{0} \approx q_S^c X^c \tag{3.4.9}$$

from which it is calculated that

$$X^c = \Phi_{\text{obs}}^{0} / q_S^c \tag{3.4.10}$$

Let us imagine now, a finite (macroscopic) and positive disturbance of the net specific metabolization rate. The relationship (3.4.9) then takes the form,

$$\Phi_{\text{obs}}^{0} = (q_S^c + \Delta q_S^c) X^c \tag{3.4.11}$$

It is well understood that this net metabolism rate increase disturbs the mass balance that is therefore no longer satisfied. Under condition (3.4.7), the only way to reestablish the matter balance is to modify the biomass. The condition of this readjustment is given by (using (3.4.11)),

$$X^c \Rightarrow X^{/c} \text{ such as } X^{/c} = \frac{\Phi_{\text{obs}}^{0}}{q_s^c + \Delta q_s^c} \tag{3.4.12}$$

$X^{/c}$ is therefore the biomass value that satisfies the mass balance when the metabolization rate increases by Δq_S^c. It is, in fact, easy to check that

$$\Phi_{\text{obs}}^{0} = q_S^{/c} X^{/c}$$

(where $q_S^{/c} = q_S^c + \Delta q_S^c$) really is an identity that satisfies the general mass balance (3.4.9).

By comparing (3.4.10) and (3.4.12), it is obvious that

$$X^{/c} < X^c \tag{3.4.13}$$

and so, under condition (3.4.7),

$$Y'_{X^c,S} < Y_{X^c,S} \tag{3.4.14}$$

In conclusion, it can be confirmed that in a situation where the quantity of substrate used remains constant (Eq. 3.4.7), all increase in specific rate of metabolism leads to a decrease in the biomass formed. This reduction is the consequence of the necessity to preserve the mass balance of the system.

This conclusion has an important corollary from the energy point of view: namely, if the increase in the specific rate of metabolism leads to a reduction in biomass at constant substrate consumption, it is evident that part of the substrate that is not changed to biomass, was used for anything else. As the formation of biomass is associated with anabolism, it can be assumed that at least one part of the substrate that is not involved with the anabolic pathway has been engaged in the catabolic pathway that produces energy and some by-products (CO_2 and H_2O). This important question will be returned to later.

Note The treatment of the most general case, using Eq. (3.4.8) without approximation, is a little complex because it includes the possibility of accumulation of free substrate in the intracellular phase. The conclusion, however, remains valid, although the exact expression of reduction in biomass (3.4.12) may be more complex. From a practical point of view, expression (3.4.12) is applicable in all cases where $\Phi_S^0 \gg D\tilde{C}_S^c$, that is to say when the net, exchange flux between the two phases is much greater than the hydraulic outlet of substrate associated with the cell phase, which is virtually always the case for easily metabolized substrates.

3.4.4 *Substrate Recycling*

Now let us consider a particular mechanism which represents the two features mentioned above, namely,

1. conservation of substrate consumed (Eq. 3.4.7),
2. associated with an increase in the net, specific rate of metabolism.

In order to make the demonstration more easily understood, let us compare two systems placed in the same experimental conditions, but which differ in the number of metabolic pathways they have.

Let us look at the diagram in Fig. 3.10a. It represents the basic configuration of the distribution of the large metabolic flows. In a way, it is the fundamental mechanism that shall be called, MEC.0. A substrate, *S*, is transported from the matrix medium to the cell phase. Next, it is distributed according to the principal fluxes: a biosynthetic flux of cellular components that will form the biomass and an

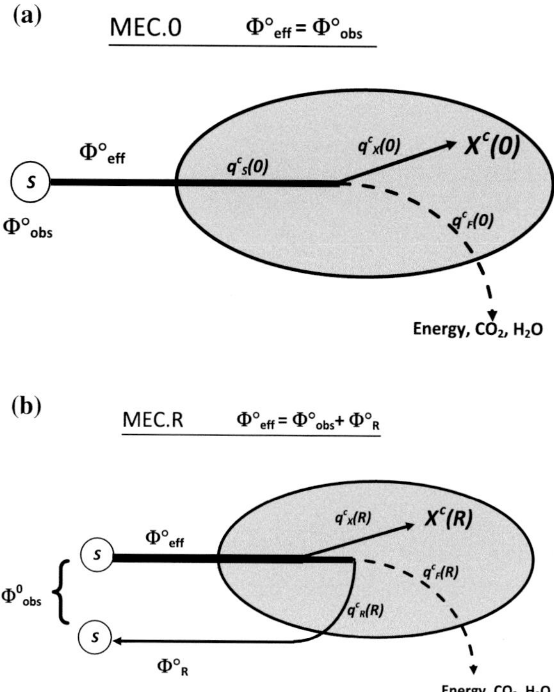

Fig. 3.10 **a** Explanation: refer to the text. **b** Explanation: refer to the text

energy production flux (or *fueling*) necessary for biosynthesis. The sum of these two flows forms the metabolic flow (anabolism and catabolism). The only products secreted following this metabolic pathway are water and carbon dioxide. This distribution of fluxes corresponds to that generally admitted, be it explicitly or implicitly (Marr 1991).

Now, let us imagine a second system corresponding to the mechanism in Fig. 3.10b, in which a supplementary flow of expulsion of substrate would come to be added (MEC.R). The two systems are placed in the same experimental conditions and it is observed that the consumption of substrate is the same in both the cases (cf. Eq. (3.4.7)). The denotations "0" and "R" will be noted in brackets in order to distinguish the variables that are related to the basic system or to the system with expulsion, respectively. So, the relationship (3.4.7) would take the form,

$$\Phi^0_{\text{obs}}(0) = \Phi^0_{\text{obs}}(R) = \Phi^0_{\text{obs}} \tag{3.4.15}$$

This indicates that the interphasic exchange flow is the same for the basic system as for the system with expulsion. In the basic system, the observable flow is the only real flux. However, in the system, R, the observable flux is a net flow, composed of an inlet term, said to be *efficient* Φ^0_{eff} and of an outlet, or release, term, Φ^0_R.

It is immediate, in the system with release, that

$$\Phi^0_{\text{eff}} = \Phi^0_{\text{obs}} + \Phi^0_R \tag{3.4.16}$$

In so far as the released substrate is identical to the substrate that is coming in, efficient flux and released fluxes are not directly distinguishable, experimentally. These are *cryptic fluxes* that require a special, experimental protocol to be shown. They cannot be shown by routine measures. The whole of the phenomena (inflow and release) constitutes the recycling of the substrate.

First of all, let us show how such a mechanism disturbs the specific substrate metabolism rate by degrading the global rate into their components. For the basic system, it is simply cf. (3.4.2)

$$q^c_S(0) = q^c_{S,F}(0) + q^c_{S,X}(0) = q^c_{S,M}(0) \tag{3.4.17}$$

$q^c_{S,F}(0)$ is the specific rate associated with fueling (or with catabolism);
$q^c_{S,X}(0)$ is the specific rate associated with the biomass synthesis (or with anabolism);
$q^c_{S,M}(0)$ is the sum of these two rates, corresponding to metabolism.

For the system with release, there will then be

$$q^c_S(R) = q^c_{S,F}(R) + q^c_{S,X}(R) + q^c_{S,R}(R) = q^c_{S,M}(R) + q^c_{S,R}(R) \tag{3.4.18}$$

where $q^c_{S,R}(R)$ is the specific rate associated with release; the other terms having the same significance as in the previous case. If the two systems present the same rate of metabolism, that is to say if

$$q^c_{S,M}(0) = q^c_{S,M}(R) \tag{3.4.19}$$

by comparing (3.4.17) and (3.4.18), it appears immediately that

$$q^c_S(R) = q^c_S(0) + q^c_{S,R}(R) \tag{3.4.20}$$

and so

$$q^c_S(R) > q^c_S(0) \tag{3.4.21}$$

that is to say, the total, specific rate of the system with release is greater than that of the reference system. Instead of considering the two different systems, 0 and R, release can be considered as a disturbance of the basic system. In this case, it is considered that the basic system is modified by the addition of a release pathway of the substrate after this has entered the cell phase. The signification of (3.4.19) is then simply that this supplementary mechanism does not modify the specific rate associated with metabolism. Recycling of the substrate is therefore a cryptic mechanism that responds to the two conditions that were mentioned at the beginning of this paragraph.

Now let us show that the biomass of system R is less than that of the basic system. The general balance (3.4.9) is to be applied for both systems. So,

$$\Phi^0_{\text{obs}} = q^c_S(0)X^c(0) \tag{3.4.22}$$

and

$$\Phi^0_{\text{obs}} = q^c_S(R)X^c(R) \tag{3.4.23}$$

By eliminating Φ^0_{obs} between these two equations and using (3.4.20), it is easily found that

$$X^c(R) = \frac{q^c_s(0)}{q^c_s(0) + q^c_{s,R}(R)} X^c(0) \tag{3.4.24}$$

and so

$$X^c(R) < X^c(0) \tag{3.4.25}$$

3.4.5 Practical Considerations

It will now be shown using a practical example that the considerations above are applicable to practical cases.

By dividing both sides of expression (3.4.22) by the biomass and using the definition of the yield coefficient, it is obtained that

$$q^c_S(0) = \frac{D}{Y(0)} \tag{3.4.26}$$

where $Y(0)$ is the yield coefficient of the biomass with respect to the substrate (cf. (3.4.6)) in the same fundamental system.

Likewise, (3.4.23) can be put in the form,

$$\frac{\Phi^0_{\text{obs}}}{X^c(R)} = q^c_S(R) \tag{3.4.27}$$

and using (3.4.20) and (3.4.27)

$$\frac{\Phi^0_{\text{obs}}}{X^c(R)} = \frac{D}{Y(0)} + q^c_{S,R}(R) \tag{3.4.28}$$

If it is assumed that $Y(0)$ and $q^{c}_{S,R}(R) = f_R$ are constants, the relationship between the specific substrate transfer flux and the dilution rate is then a straight line of slope, $1/Y(0)$ and the intercept at the origin is $f_R > 0$. So the characteristic relationship of the systems said to have "nonzero maintenance energy" is found (Pirt 1965; Tsai and Lee 1990). Obviously, if the specific release rate is zero, (3.4.28) is a straight line passing through the origin ("maintenance energy is zero"). So there is an unexpected parallelism between the specific release rate and the maintenance energy concept.

To be absolutely certain, let us examine the dependence of biomass according to the dilution rate by transforming (3.4.28), it is easily obtained that

$$X^c(R) = \frac{D\left(\tilde{C}_S^{m,E} - \tilde{C}_S^{m}\right)}{D/Y(0) + f_R} \tag{3.4.29}$$

So the usual hyperbolic D relationship, linking biomass, dilution rate and the limiting substrate (Nielsen and Villadsen 1994) is found again. In the relationship (3.4.29), $Y(0)$ plays the role of the "true yield coefficient," or $Y^{\max}$, but here the concept of maintenance coefficient is replaced by that of the specific release rate, f_R. Figure 3.11 shows the adjustment of (3.4.29) according to the dilution rate.

The experimental data are those of Nielsen and Villadsen (1994) for the steady states of *Enterobacter* (*Aerobacter*) *aerogenes* cultivated in a chemostat. At low

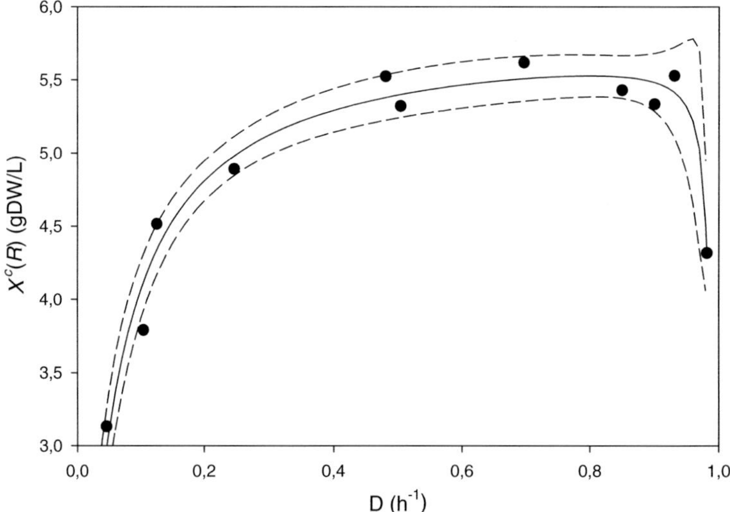

Fig. 3.11 Profile of biomass (dry weight) in the system with release, according to the dilution rate. The experimental data (*black circles*) are that of Nielsen and Villadsen (1994). The adjustment was made using relationships (3.4.29) and (3.4.30). Parametric estimation gives $a = 0.022$ g L^{-1}; b = 1.00 g L^{-1}; $Y(0) = 0.59$; and $f_R = 0.077$ h^{-1}. The curves in *dotted lines* represent the 95 % confidence interval

cell densities the pseudo-homogeneous concentration little differs from the measured concentration. The error caused using the experimental values without corrections is very much less than measurement errors. So "raw" experimental values have been used. For the residual substrate determination, the following relationship was used:

$$\tilde{C}_S^m = \frac{aD}{b-D}, \quad \tilde{C}_S^m < \tilde{C}_S^{m,E} \tag{3.4.30a}$$

$$\tilde{C}_S^m = \tilde{C}_S^{m,E}, \quad \tilde{C}_S^m \geq \tilde{C}_S^{m,E} \tag{3.4.30b}$$

Parametric estimate of residual substrate gave $a = 0.024$ g/L and $b = 0.99$ g/L ($r^2 = 0.996$), indicating that the function (3.4.30) is adequate. It is, however, possible, by introducing (3.4.30) into (3.4.29), to proceed to the parameter estimate in one step. The result obtained was then $a = 0.022$; $b = 1.00$. The "real" maintenance coefficient is then $Y(0) = 0.59 \pm 0.01$ and the specific release rate, $f_R = 0.077 \pm 0.007\ \mathrm{h}^{-1}$. The correlation coefficient for the determination of the four parameters is a little less, $r^2 = 0.965$. Adjustment of (3.4.29) to experimental data can be considered as very satisfactory.

3.4.6 Discussion

Using the definition of the yield coefficient in (3.4.28), it is obtained that

$$D\left(\frac{1}{Y(R)} - \frac{1}{Y(0)}\right) = q_{S,R}^c(R) > 0 \tag{3.4.31}$$

from which it is immediately found that

$$Y(R) < Y(0) \tag{3.4.32}$$

This means that recycling of the substrate, superimposed on anabolic and catabolic pathways, leads to a lower biomass to substrate yield. On the other hand, the relationship (3.4.19) implies that specific metabolic rate remains invariable and so that recycling is a distinct metabolic pathway. Decrease in yield (3.4.32) must therefore be explained by the fact that part of the energy from the catabolic pathway is used to the advantage of the substrate recycling. In other terms, recycling of the substrate must be an active process and consume energy at the expense of biosynthesis. Three mechanisms could possibly explain recycling consumption energy:

(a) Transport is active. This is obviously a likely hypothesis. In our case, the energy spent on transport evidently would be increased even more by the fact that the transport flux is increased (Eq. (3.4.16)). The extra energy loss in relation to the basic system would then be due to this increase in flux.

(b) Substrate release is active. A phenomenon of excretion of substrate that would consume energy directly to expel the compound cannot be thought out of the question. This may be done indirectly, by synthesis of specific excretions or by other phenomena of membrane polarization. Experimentally, the possibility of energy dissipation by a phenomenon of release (following a metabolic overflow) was pointed out by Streekstra et al. (1987).
(c) Finally, the possibility that the substrate is partially degraded and resynthezised during the process needing energy, cannot be excluded.

The model does not make it possible to favor one hypothesis rather than another and does not exclude a combination of these various mechanisms. The necessary and sufficient condition for the relationship (3.4.32) to be possible, is that the net energy balance of the process is not unfavorable to the system. The conclusion here is in agreement with the mosaic nonequilibrium thermodynamics (MNET; Westerhoff et al. 1982).

Moreover, expulsion of the substrate in continuous culture was shown experimentally in several systems by the isotopic, relaxation method (Button and Kinney 1978; Button 1991). Using this method, Robertson and Button (1979) did a very definitive study of *Rhodotorula* (*rubra*) *mucilaginosa* cultivated in chemostat with phosphorous as limiting substrate. The rate of substrate release was proved near constant and the ratio of rate of expulsion over the rate of uptake was decreasing with D. Adjustment of their data made possible to estimate that the ratio was from about 11 % for 0.25 μ_{max} and around 2 % for 0.75 μ_{max}. These values are close to those found for *Klebsiella pneumoniae*. Let us also point out that a mechanism for simultaneously collecting and expulsion of substrate was observed in *Streptomyces* (*hydrogenans*) *exfoliatus* growing in chemostat in a carbon-containing medium (glucose and succinate), enriched with amino acids (Alim and Ring 1976). In this study, it appears that the inflow as well as the outflow of cycloleucine is growth rate dependent. Release and reuse of ethanol in some yeasts is also a well-known phenomenon (Rieger et al. 1983) that argues in favor of recirculation of certain compounds.

The proposed mechanism of substrate recycling is hence realistic, only the cryptic nature of the fluxes has, doubtless, made it possible for this phenomenon to occur unnoticed.

However, it is not very likely that substrate recycling is the only explanation for the use of energy for other purposes than growth. Surely other causes play their parts. Maintenance properly called, (turnover of proteins, cell motility, etc.) or endogenous metabolism (futile cycles) certainly uses energy at the expense of growth. Recently, Bond and Russell (1998, 2000) explained energy waste by ATPase in *Streptoccocus bovis*. However, it is a question of homolactic fermentation that does not correspond to the physiological mechanism of Fig. 3.10. The study of the case that appears in (3.4.2) shows that substrate recycling makes possible interpretation of all dissipation of energy. In the case of strictly aerobic respiration with water and carbon dioxide as the only final products, the hypothesis that this mechanism predominates maintenance and endogenous metabolism, can be made. Dissipation of energy is in fact, indispensable to the good cell functioning for thermodynamic reasons (Westerhoff et al. 1982). In the situation described in Fig. 3.10, substrate recycling is perhaps one of the only alternatives good enough to satisfy this thermodynamic constraint. Supplementary works are, without doubt, necessary to evaluate case by case, the proportion relative to the various phenomena that play a significant part in the consumption of energy not used in growth. As emphasized in the introduction, some industrial processes are very sensitive to slight increase in the efficiency of a process. Even a marginal, energy waste phenomenon can be important.

To conclude, attention is drawn to the perspectives that open up the substrate recycling mechanism to improve industrial strains of microorganisms. The relationship (3.4.21) can be interpreted as the optimization of net, specific rates of phenomena of transport/metabolism when the substrate is limited. On the contrary, this time, in line with the predictions of mosaic, thermal nonequilibrium (Westerhoff et al. 1983), the optimization of these rates is done at the expense of biomass and favors dissipation of energy according to biosynthesis. (Note that the diagram in Fig. 3.10b has not been considered in MNET and that the conclusions concerning the optimization of growth of biomass are obtained from other metabolic diagrams. An explanation is beyond the scope of this discussion.) In the case where relationship (3.4.21) is applicable, there is hope that there could be an improvement in the efficiency in biomass production by reducing (or by inhibiting) the rate of transport of substrate in the cell. This approach is not intuitive and could lead to unexpected results. The other important way for improvement in strains is a return to the active mechanisms of excretion expected from the model. However, as far as it is known, these mechanisms are still largely unknown as far as the excretion of carbonated substrates is concerned (contrary to what happens with excretion of proteins).

3.4.7 *Additional Considerations*

The results obtained in this section are very fruitful and open up numerous perspectives. Moreover, it has been noted that the recirculation of substrate and its implications as far as energy is concerned, were not always intuitive. For these two reasons, a series of reflections and speculations are added to the conclusion.

3.4.7.1 A Hydraulic Analogy

The expulsion mechanism followed by uptake could be compared to a hydraulic system for recirculation. A pump $P1$ (cf. Fig. 3.12) makes possible the inflow of a liquid into a reservoir R with a volumetric flow Q_{in}. At the outlet of the reservoir, a fraction of the flow is recycled to the inlet of the reservoir, thanks to a pump $P2$, with a volumetric flow Q_{out}. The mass balance around the whole of the system

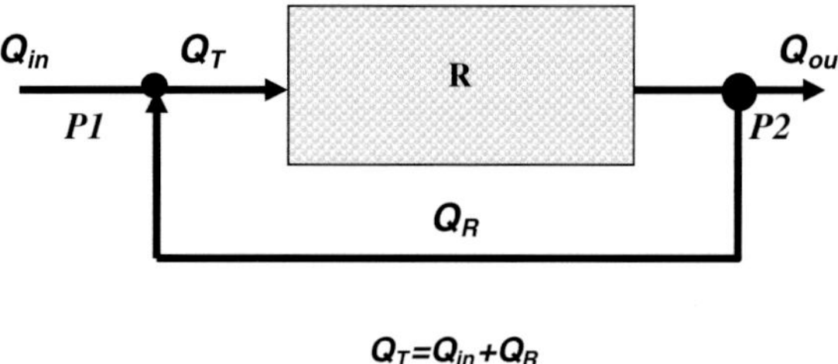

Fig. 3.12 Diagram of the hydraulic analogy for the model with expulsion. Although the real mass flow of the liquid drawn from outside is Q_{in}, the inflow itself in reservoir R is greater and $Q_T = Q_{\text{in}} + Q_R$. The global balance remains invariable and $Q_{\text{in}} = Q_{\text{out}}$. The only gain with the system without recycling is an increase in inflow. Pump $P2$ decreases the energy yield of the system with recycling

shows that $Q_{\text{in}} = Q_{\text{out}}$. This result is obviously the same for a system without recycling. So what is the difference between a system with recycling and a system without recycling? First, the inflow into the reservoir is increased by recirculation and $Q_T = Q_{\text{in}} + Q_R$. So, for the same quantity of liquid available outside and without consuming more, it is possible to increase the inflow. On the other hand, there is a price to be paid to achieve this. Pump $P2$ (or all other mechanisms) is necessary for recycling and consumes energy. In other terms, the increase in inflow in the reservoir implies a decrease in the global energy yield of the system.

3.4.7.2 Biological Interpretation

What is to be gained by spending energy on conserving a higher, partial, specific rate of uptake? At this stage, it is mere speculation. For example, it might be conceived that energy loss due to expulsion is compensated by better access to the outside substrate. Or even that the cell maintains functioning uptake apparatus well, in case of famine that is immediately efficient when more substrate becomes available. Another class of hypothesis would make the expulsion a form of cell regulation destined to maintain the intracellular concentration of the substrate below a critical threshold that is not harmful to the cell. The release energy would then be devoted to detoxification due to excess substrate. It could even be speculated that there is “altruistic” behavior on the part of the microorganisms. In a poor substrate medium, the release induces a biomass decrease. Consequently, the biomass/substrate ratio available becomes more favorable to the whole population. Release would then take the form of a redistribution of resources in order to maintain a smaller but more viable population. There are more hypothesis of a thermodynamic nature. For example, it is admitted that microbial growth on substrates that are greatly reduced imposes a dissipation of the excess of free enthalpy. Loss in effectiveness of this type of growth implies a phenomenon of dissipation decoupled from biomass production (Westerhoff et al. 1982). The release model and the associated spilling of energy corresponds well to this requirement. It would then be a purely dissipative process destined to satisfy the constraints of a thermodynamic nature. Other studies will be necessary to improve the understanding of the phenomenon, but the preceding hypotheses illustrate well, that the expulsion model can rest on reasonable biological justifications and do not constitute an ad hoc solution to the explanation of certain experimental phenomena. A final argument will be developed in the paragraph that follows.

3.4.7.3 Variability in Specific Maintenance Rate

The specific maintenance rate is sometimes defined by $a = Y_G \cdot m$, where Y_G is the maximum yield from growth coefficient, and m the maintenance coefficient (Pirt 1982). In his article, Pirt raises the following question, "Why should the specific maintenance rate (a) vary up to 30 times (about 0.01 h^{-1} up to 0.3 h^{-1}) according to the nature of the energy source or carbon source that limits growth?" The same type of question occurs regarding the variability of the maximum yield coefficient for ATP (Y_{ATP}^{max}; Westerhoff et al. 1982). It is well known that the composition and the physiological state of the cells of a microorganism depend on its growth rate and not (or nearly not at all) on the nature of the substrate on which it grows (Neidhardt et al. 1995). Knowing this, Pirt's question might be rephrased in the following manner: "So why when the composition and the physiological state of cells are independent of the nature of the substrate, does the specific maintenance rate depend on the nature of the substrate?" Within the perspective of the model, the "specific release rate" would be expressed by $a = Y(0) \cdot f_R$ (with $Y(0) \equiv Y_G$). It is in no way surprising that the release mechanism of the substrate varies according to the nature of the expulsion. It is even more natural to think that the release coefficient, f_R, depends on the nature of the substrate. The same goes for the release energy. Each substrate can therefore be characterized by a characteristic release and energy coefficient. If the notion of physiological state is defined according to cell metabolism, as is the case, the decoupling between metabolism and release makes it possible to reconcile the notion of a unique, physiological state associated with different mechanisms expulsion. Variability in specific release rate and variability in the yield coefficient for ATP according to the nature of the substrate (and/or the microbial species) ceases to be problematic within the perspective of the release model.

3.4.8 Should the Concept of Maintenance Be Abandoned?

This question probably requires a long debate that is only touched on here. It is enough to open up a dialogue with a definite example. In the numeric example, the constant value that appears in (3.4.29) can be replaced by a value of the type $k = f_R + m$ = constant that would associate expulsion (f_R) and maintenance (m). In a chemostat, however, the cells can be considered as undergoing a constant, exponential growth, implying an increase in size and incessant cell division. This situation can be contrasted with the state of cells cultivated in batch in a quasi-steady state (*resting cells*). In this last case, the energy that may be consumed is used to preserve a constant cell structure (without reproduction or increase in size). In this case, maintenance energy can be spoken of without ambiguity. For active cells cultivated in chemostat, it is thought more reasonable to assume that the maintenance energy is negligible compared with the total cell energy balance and so $k \approx f_R$. Finally, it is just conceivable that the two concepts be preserved in a mutual way, but it seems clear that maintenance alone is far from accounting for chemostat phenomena. (In support of this idea, let us recall that the theory of maintenance predicts that in a steady state, a limited (or maximum) biomass would be reached in a chemostat that is 100 % recycled. Chesbro et al. (1979) demonstrated experimentally that this prediction was not right). Of course, the relative importance of the two phenomena remains to be determined; nevertheless, it might be more advantageous to give precedence to the expulsion model and include the marginal energy phenomena (such as the futile cycles or membranous polarization) in an enlarged form of the yield coefficient and to consider them as normal, metabolic events.

3.5 Threshold Phenomena, Signals in Cells, and Metabolic Pathway Switches

3.5.1 Notice

This section is extremely theoretical and can, if absolutely necessary, be omitted without seriously harming the unity and understanding of the text. The notions presented here will, moreover, be taken up again in summarized form in the sections where they are needed. So this part could have been taken out and annexed or "distilled" with abstemiousness in the parts where needed. However, it was preferable to place this section here, which is its logical place in the structure of the book, for the following reasons:

- the progressive development of the various models that follow really helps ingood understanding of the two sections that follow;
- changes (that are chronological) in the various models, show the logic that has led to the development of the most complex model used in the rest of the text;
- it is also preferable to become aware of the theoretical details before embarking on applications where these are reduced to a summary, in order to identify the biological meaning rather than concentrating on the mathematical features.

Instead of purely and simply omitting this section, a concise résumé of the various models described is given. This will more or less help a better understanding of what follows and will also make it possible to take into consideration the perspectives and new applications that perhaps justify a complete reading of this section.

3.5.2 Summary

A series of explicit forms of the general implicit balance obtained in Chap. 2, are presented here while limiting the study to the biphasic chemostat as defined in Sect. 3.3.

By explaining the specific intracellular transport and metabolism (T/M) rate in the form

$$q_S^c = \sum_{i=1}^{n} \frac{V_S^0(i) C_S^c}{K_S(i) + C_S^c}$$

a new class of models (theoretically infinite) is created in which all intracellular kinetics are hyperbolic with respect to the concentration of the given compound, S. In the rest of the text, just systems with one (n = 1) or two (n = 2) pathways are considered.

3.5.3 Model with One Pathway *n* = 1

- **Case 1**: Very high maximum T/M rate ($V_S^0(1) > D_W$)

  The properties of this model are as follows:

- progressive, but slight accumulation of intracellular compound according to D;
- great sensitivity to concentration S according to the affinity ($1/K_S$) of the T/M for the compound;
- very high specific rate (near to the maximum), without saturation and weak sensitivity to affinity.

Discussion This situation rather corresponds to a simple and essentially continuous but very effective regime for transport and metabolization of the compound. The contribution of the global growth rate ("μ") can be high, especially if the compound is used in the anabolic pathway. On the whole, this model describes a cell that is very well equipped to treat the compound S; however, it must fulfill the constraining condition for this which imposes that the T/M rate is greater than washout. Apart from this, one of the weaknesses of the system resides in the strong limitation of means to regulate the intracellular concentration.

– **Case 2**: Maximum T/M rate lower ($V_S^0(1) < D_W$)

The properties of this model are as follows:

- accumulation of intracellular substrate is large, progressive (with weak affinity), or more sudden (with strong affinity);
- very rapid, decreasing sensitivity of concentration of S to affinity;
- higher specific rate below the threshold for the dilution rate, but with saturation above this threshold.

Discussion At weak affinities and low dilution rates, cell behavior is very near the preceding case. At high affinities, threshold phenomena can appear. This representation is convenient to explain the sudden accumulation of intracellular compound above a certain growth rate (threshold). This sudden rise in intracellular S can possibly lead to secondary effects and serve as a chemical signal associated with rate and/or concentration of the compound in the matrix phase (concentration also depends on the rate). The cell capacity is limited because the T/M rate saturates at (or after) the critical value, and this occurs whatever the value of affinity for the compound. It is this saturation that makes accumulation of the compound possible. Note that too great a simplification of the system prevents all forms of precise regulation.

3.5.4 Model with Two Pathways $n = 2$

The only model with two pathways that will be considered includes one pathway with high affinity and one pathway with low affinity. Let us consider two situations:

– **Case 1**: Model without metabolite excretion.

The model is then adapted to the representation of storage phenomena with threshold effect and possibly applies to the description of phenomena that make chemical signals appear. The pathway with low affinity intrudes greater flexibility in the regulation of the intracellular concentration both in a structural way (via the constant kinetic value representing the ratio of the maximum T/M rate and affinity S) and through capacity of the model to introduce a metabolic switch by "turning on" the pathway with low affinity beyond the critical dilution rate. So complete saturation of T/M that is characteristic of the one pathway model, is avoided.

– **Case 2**: Model with excretion of metabolite.

The properties of this model include those of the preceding case. Excretion of metabolite makes it possible, however, for new properties to appear, such a threat of regulation of the metabolism is by adjustment of the biomass and regulation of the total specific rate. This situation, as fruitful as it is interesting is abundantly treated in Sect. 3.6 (Table 3.6).

Table 3.6 Summary table of explicit representations

Case	Characteristics of the model	Cell effect
$n = 1$ V_S^0 high	Only one T/M pathway with raised maximum rate	Physiological treatment of S very efficient. Regulation of intracellular concentration very limited
$n = 1$ V_S^0 small	Only one T/M pathway with low maximum rate	Good capacity for storing S with or without threshold effect. Adapted to the description of chemical signals. Regulation of the intracellular concentration very limited
$n = 2$ Excretion = 0	Two T/M pathways – one pathway with low affinity – one pathway with high affinity No excretion of metabolites	Good storage capacity with threshold effect. Adapted to the description of chemical signals. Good regulation of the intracellular concentration. Metabolic switch capacity
$n = 2$ Excretion $\neq 0$	Two T/M pathways – one low affinity pathway – one high-affinity pathway Excretion of metabolites	The properties of the model are comparable to the preceding case, but also take into account excretion of a metabolite and a phenomena of regulation by adaptation of the biomass

3.5.5 Introduction

Basing this on Bellgardt's analysis (1991), nonstructured models that present a global description of phenomena (on the global scale of the bioreactor) are limited as soon as it is a question of having to take into account a dynamic at cell level. The structured models that take into account certain facts concerning the physiological state of the cell (and which give details, in this way, of a more local description) have made it possible to represent more complex phenomena. However, they remain quite vague as regards intracellular dynamics and, in the case of structured models with compartments, the complexity of the models increases in a prohibitive way, according to the number of compartments, which limits working these on a practical level. To take into account mainly the growth phenomena on media of multiple substrates, models called cybernetic (Cortassa et al. 2002; Varner and Ramkrishna 1999) or "with a metabolic regulator" appeared in the 1980s. These models focus much more on a local description of the intracellular kinetics and have proved very effective in representing certain complex phenomena. The principle objection to these last types of modeling concern their distance from regulation mechanisms and use very generous concepts of optimization (such as maximization of growth rate, for example) which may not be legitimate (Varma and Palsson 1993a, b). The question of finding out how regulation takes place is not asked and optimization rests on the fact that it is the biological development that brings choice of optimum process (Bellgardt 1991). Technically speaking, the consequences of this conceptualization imply that at a given moment of the working of the kinetics of the system, several options can be presented of which the model maker gave in advance, in his algorithm, the choice of option that corresponds best to the criteria of preestablished optimization.

One of the major phenomena that has led to the development of this type of modeling is growth on multiple substrates that presents a diauxie phenomena. The problem that presents is then of correctly representing the switch that makes it possible to pass from one growth that occurs first on its own on the preferred substrate, to the consumption of the second, when the first is exhausted. The models with compartments can take into account phenomena of this kind because changes are sufficiently slight or occur gently; nonstructured, cybernetic models with metabolic regulator use strategies for choice of metabolic pathways that rest on discontinuous functions or on criteria for optimization that are both defined a priori.

The model that is presented here is a new approach to take into account switch phenomena between various metabolic pathways by coping without both the discontinuous functions and the criteria for optimization. It is based on the fact that certain realistic kinetics present the intrinsic property of activating or not, the different metabolic pathways in response to the constraints imposed by the conditions of culture, without a single previous ad hoc condition being necessary. The possible transitions can be "gentle" (or "smooth") but also extremely abrupt and take into account threshold phenomena. The model results from concepts near to the analysis for metabolic flows (*MFA*; cf. for example, Schügerl and Bellgardt 2000) with the added advantage that it is not necessary to know the biochemical reactions of the metabolism in detail. This method consists of making the transport/metabolism rates explicit. These were first obtained in an implicit form within the scope of polyphasic, dispersed systems in general (refer to Chap. 2). Description at the level of culture is chosen since this is situated between the global level of the reactor and the level of intracellular reactions. Consequently, the explicit form of transport/metabolic reactions makes it possible to calculate and represent the main specific rates of metabolism that occur in the cell phase.

3.5.6 Implicit Mass Balance

Use is made of operating conditions of the biphasic chemostat, such as those described in Sect. 3.3.

3.5.6.1 Cell Phase

Let S be a compound that can be metabolized. In a biphasic chemostat, the mass balance of compound S, transported from the dispersing matrix phase m toward the micellian phase c (cell), which is written in terms of pseudo-homogeneous concentrations (refer to (3.3.12a)),

$$\frac{d\tilde{C}_S^c}{dt} = -D\tilde{C}_S^c + \Phi_{S,m}^0(c) - q_S^c(.)X^c + \tilde{C}_S^c \frac{d \ln N_T^c}{dt} \tag{3.5.1}$$

where

$D = Q/V_T$ is the dilution rate;
$\tilde{C}_S^c$ is the pseudo-homogeneous concentration of the compound in the cell phase;
$\Phi_{S,m}^0(c)$ is the interphasic, global exchange flow (in the direction $m \rightarrow c$) per unit of volume;
$q_S^c(.)$ is the specific transport/metabolism rate of the compound in the cell phase; (the denotation (.) indicates that this is a complex function, possibly depending on numerous factors. This parenthesis is often omitted afterwards.)
X^c is biomass;
N_T^c is the total number of cells.

So the steady state of (3.5.1) is

$$\Phi_{S,m}^0(c) = q_S^c(.)X^c + D\tilde{C}_S^c \tag{3.5.2}$$

that expresses that the interphasic, transfer flow per unit of volume is equal to the sum of the net specific metabolism rate of the compound, multiplied by biomass and by the outlet term of the chemostat. This last term expresses the outlet of the intracellular compound associated with the hydraulic outlet of cells.

The relationship (3.5.2) can be put in the form (with $q_S^c \equiv q_S^c(.)$)

$$q_S^c = \frac{\Phi_{S,m}^0(c)}{X^c} - \frac{D\tilde{C}_S^c}{X^c} \tag{3.5.3}$$

Using the usual definition (that is to say, by ignoring maintenance) of the yield coefficient ratio with respect to S:

$$Y_{X^c,S} = \frac{X^c}{\tilde{C}_S^{m,E} - \tilde{C}_S^m} \tag{3.5.4}$$

and the mass fraction of the compound (mass of the intracellular compound per unit of biomass; cf. (3.3.17))

$$\alpha_S^c = \frac{\tilde{C}_S^c}{X^c} \tag{3.5.5}$$

is derived (refer to Table 3.4)

$$q_S^c = D\left(\frac{1}{Y_{X^c,S}} - \alpha_S^c\right) \tag{3.5.6}$$

This relationship always holds true for a transport phenomenon followed by consumption. The relationships (3.5.3) and (3.5.6) express that the effective transport/metabolism rate is the difference between the interphasic, specific transport flow and the outlet of compound associated with the cells leaving the reactor. Effective transport/metabolism rate is therefore not directly observable and can only be calculated if the intracellular concentration is known.

It is evident that

$$0 \le \alpha_S^c < \alpha_{S,\max}^c \tag{3.5.7}$$

Theoretically, the maximum value of $\alpha_{S,\max}^c$ is 1. However, this means that the entire cell phase is formed only of intracellular compound, which is obviously absurd. (Realistic values for $\alpha_{S,\max}^c$ can, however, sometimes be very much raised as regards compounds that are difficult to metabolize or when storing takes place.) The value 1 therefore constitutes an increasing absolute value (theoretical limit).

Extreme values (minimum and maximum) of the metabolic rate are therefore determined by the limits (3.5.7) from which

$$\min(q_S^c) = D\left(\frac{1}{Y_{X^c,S}} - \alpha_{S,\max}^c\right) \tag{3.5.8}$$

and ($\alpha_S^c = 0$; completely metabolized compound)

$$\max(q_S^c) = \frac{D}{Y_{X^c,S}} \tag{3.5.9}$$

This last relationship is very important when the compound S is the limiting substrate.

Note When the intracellular concentration is very small compared with the inverse of the yield coefficient, the relationship (3.5.3) shows that the specific, transfer flow is equal to the rate at which the compound disappears. In this case, the relationship (3.5.9) shows that this consumption rate varying according to dilution rate, is a straight line passing through the origin and of slope equal to the inverse of the yield coefficient. Experimentally, if S is the limiting substrate, this is not always the case and a certain number of cultures show a straight line that does not pass through the origin. This consumption of substrate, associated with no growth ($D = 0$), is generally interpreted in terms of cell maintenance energy (Pirt 1965, 1982; Tempest and Neijssel 1984). This problem has been abundantly treated in the preceding section (Sect. 3.4).

3.5.6.2 Matrix Phase

In the dispersing matrix phase, in absence of reaction of S in the matrix phase, the mass balance of the compound is written (refer to (3.3.12b)) as

$$\frac{\mathrm{d}\tilde{C}_S^m}{\mathrm{d}t} = D\left(\tilde{C}_S^{m,E} - \tilde{C}_S^m\right) - \Phi_{S,m}^0(c) \tag{3.5.10}$$

where $\tilde{C}_S^{m,E}$ and $\tilde{C}_S^m$ are concentrations of the compound in the matrix phase at the inlet and in the body of the reactor, respectively. In the steady state

$$\Phi_{S,m}^0(c) = D\left(\tilde{C}_S^{m,E} - \tilde{C}_S^m\right) \tag{3.5.11}$$

which expresses that the flow transferred to the cell phase, is simply the difference between the inflow and the outflow of the bioreactor.

3.5.7 *Explanation of Specific Rate*

Results obtained up to now are general, implicit results already gained. Now let us make the specific rate explicit by introducing some simple, minimalist hypotheses.

3.5.7.1 Transport/Metabolic Kinetics (T/M)

The net specific, global rate that finally makes metabolism of the compound possible in the cell is the result of several processes,

(a) diffusion of the compound from the matrix phase toward the membrane;
(b) transport of the compound from the exterior to the interior of the cell;
(c) metabolism (complete or incomplete) of the compound in the cell.

We will call transport/metabolism (T/M) the rate resulting from these three processes.

Diffusion of the substrate in the cell membrane leads to the expression of specific T/M rate in terms of local concentration (Coulson and Richardson 1987), that is to say in terms of "reacting concentration" (or R-concentration, refer to Sect. 2.3). The general expression of this specific rate is therefore in the form

$$q_S^c = f(C_S^c, \ldots) \tag{3.5.12}$$

where C_S^c is an R-concentration.

As net rate, this form can always be broken up into a series of terms that take into account the multiplicity of the transport pathways and the diversity of the metabolic pathways,

$$q_S^c = \sum_i f_i(C_S^c, \ldots) \tag{3.5.13}$$

An explicit form of (3.5.13) can be obtained using a very general hyperbolic function to represent the specific T/M rate,

$$q_S^c = \sum_{i=1}^{n} \frac{V_S^0(i) C_S^c}{K_S(i) + C_S^c} \tag{3.5.14}$$

where $V_S^0(i)$ is the maximum T/M rate for pathway i and $K_S(i)$ is the affinity for the compound corresponding to this pathway (affinity is in fact the inverse of the constant). This hyperbolic relationship is very often used in this context and a similar kinetic has already been used by Weusthuis et al. (1994) for the description of transport phenomena by multiple transporters. Under consideration will be the case of $n = 1$ (model with one pathway) and the case of $n = 2$ (model with two pathways).

3.5.7.2 *n* = 1—Model with One Pathway

The relationship between *R*- and *E*-concentrations (refer to (3.3.17)) is given by

$$C_S^c = \tilde{C}_S^c \frac{\delta_c}{X^c} \tag{3.5.15}$$

where δ_c is the density (g/L) of the cell phase.

By applying this relationship to (3.5.14) and for $n = 1$, it is easy to show that

$$q_S^c = \frac{V_S^0 \tilde{C}_S^c}{K_S^* X^c + \tilde{C}_S^c} \tag{3.5.16}$$

where $V_S^0 = V_S^0(1)$ and $K_S^* = K_S(1)/\delta_c$ (note that K_S^* has no units in this theory).

The special feature of (3.5.16) is to show that global affinity ($K_S^* X^c$) is no longer constant but depends on biomass. This property has been reported by several authors and justified within the scope of the collisional limit (Abott and Nelsestuen 1988). This point will be returned to in the discussion.

Using the form (3.5.16) in the implicit mass balance (3.5.2), the stationary state is given by

$$\Phi_S^0 = \frac{V_S^0 \tilde{C}_S^c}{K_S^* X^c + \tilde{C}_S^c} X^c + D\tilde{C}_S^c \tag{3.5.17}$$

where, in order to simplify writing the equation, it is noted that $\Phi_S^0 \equiv \Phi_{S,m}^0(c)$.

Φ_S^0, X^c, and D are easily measurable data; $\tilde{C}_S^c$ is the pseudo-homogeneous concentration of compound S that can be calculated at the steady state thanks to

(3.5.17) to satisfy the mass balance. This expression can be put in polynomial form with variable coefficients,

$$P^2(\tilde{C}_S^c) \equiv a_2\left(\tilde{C}_S^c\right)^2 + a_1\tilde{C}_S^c + a_0 = 0 \tag{3.5.18}$$

with

$$\begin{aligned} a_0 &= -K_S^* X^c \Phi_S^0 \\ a_1 &= X^c\left(V_S^0 + DK_s^*\right) - \Phi_S^0 \\ a_2 &= D \end{aligned} \tag{3.5.19}$$

It can be shown that (3.5.18) has only one real, nonnegative root given by

$$\tilde{C}_S^c = \frac{-a_1 + \sqrt{a_1^2 - 4a_2a_0}}{2a_2} \tag{3.5.20}$$

and a zero solution for the trivial cases $X^c = 0$ and $\Phi_S^0 = 0$. (The case $K_S^* = 0$ is considered as not relevant at this stage.)

In order to study the properties of (3.5.18), either experimental data must be available or use made of a "generating model" that generates data. This latter solution will be used, in the form of Monod's model (refer to Sect. 2.2), that makes it possible to generate the data in the following manner. In the steady state,

$$S = \frac{K_M D}{\mu_{\max} - D} \tag{3.5.21a}$$

$$X = Y_M(S^0 - S) \tag{3.5.21b}$$

Use is made here of Y_M to show that it is a question of a quantity associated with the generating model (Monod's).

These relationships are valid on condition that D is less than washout rate,

$$D_W = \frac{\mu_{\max} S^0}{K_M + S^0} \tag{3.5.21c}$$

for $D > D_W$, $X = 0$ and $S = S^0$. (S, S^0: substrate in the reactor and at the inlet; X: biomass; Y_M: yield coefficient; $\mu_{\max}$: maximum growth rate; K_M: half-saturation constant.) On condition that the cell density is not too high, the following approximations can be adopted: $X^c \approx X$, $\tilde{C}_s^m \approx S$, and $\tilde{C}_s^{m,E} \approx S^0$.

Using the relationships (3.5.2), it is possible to calculate the coefficients (3.5.19) and to test the properties of (3.5.18). It appears that the two types of situation that are qualitatively different can be presented according to the value of V_S^0.

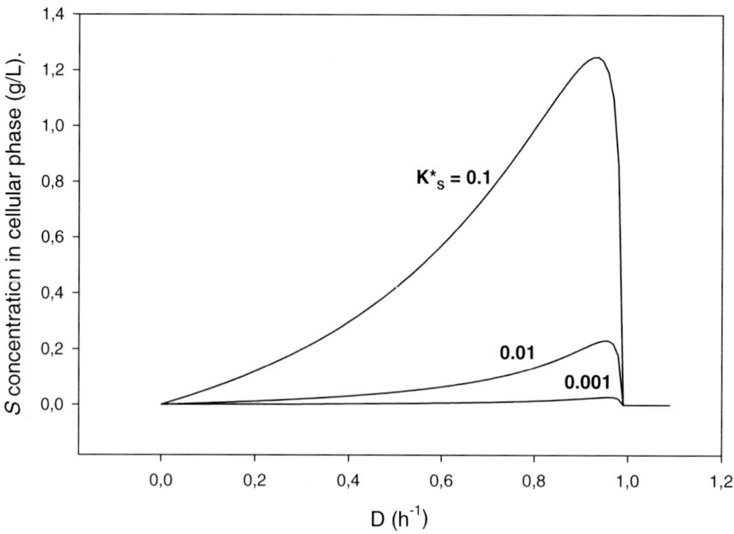

Fig. 3.13 Pseudo-homogeneous concentration variation of the compound in the cell phase according to the dilution rate. Explanations: refer to the text. Parameters of the generating model (Monod): $\mu_{max} = 1$; $K_M = 0.1$; $Y_M = 0.55$; $S^0 = 10$ g/L. PDS parameter: $V_S^0 = 2$ (h^{-1}); $K_S^* = 0.1$, 0.01, 0.001 (refer to curves)

$$V_S^0 > D_W \text{ (}\textit{High specific } T/M \textit{ rate}\text{)}$$

Figure 3.13 shows the profile of the compound in the cell phase. (The model does not make it possible to tell whether the compound is adsorbed on the membrane or if it penetrated into the cells. However, conceptually, the compound S must be associated with the solid phase and must have left the matrix phase.)

A continuous increase in the compound is observed up to a suddenly decreasing value for D. Over the whole of the domain of D, it is mainly the value of K_S^* that determines the order of size. The concentration in the cell phase decreases when affinity ($\cong 1/K_S^*$) is increased. When K_S^* tends to zero, the concentration also tends to zero.

Figure 3.14 shows that the net specific rate, q_S^c, tends toward its maximum value when affinity tends to infinity ($K_S^* = 0$). Note that using (3.5.9), it can be derived that

$$\max(q_S^c) = \frac{\Phi_S^0}{X^c} = \frac{D}{Y_{X^c,S}} \equiv \frac{D}{Y_M} \tag{3.5.22}$$

In other terms, the maximum rate of the system tends toward that of the generating model. However, note that Y_M is only an empirical parameter destined to simulate experimental values. Inside the system which is described, there is no precise signification; it is only a special value for $Y_{X^c,S}$, the "true" yield coefficient of the system.

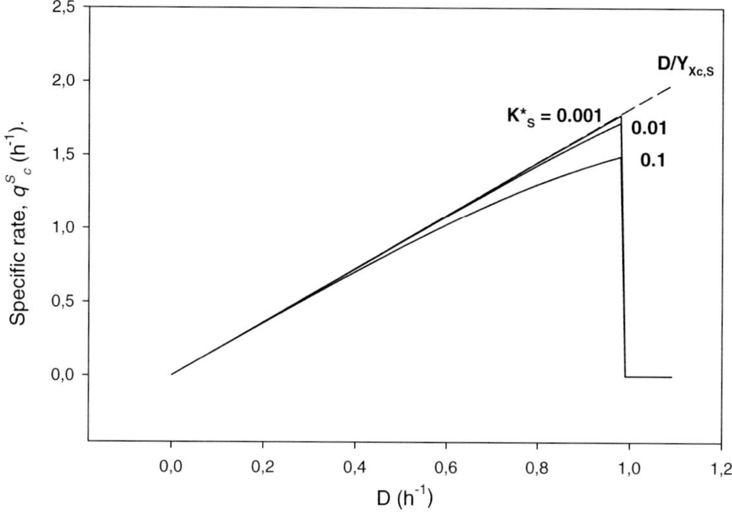

Fig. 3.14 Specific transport/metabolic rate variation according to the dilution rate. Explanations: refer to the text. Parameters of the generating model (Monod): as in Fig. 3.13. PDS parameters: as in Fig. 3.15

$$V_S^0 < D_W \text{ (Low specific } T/M \text{ rate)}$$

When the maximum specific T/M rate goes below the washout value, the situation changes completely. Figure 3.15 shows that the variation in concentration of the compound is increased according to the preceding situation (cf. Fig. 3.13).

Moreover, when the affinity is increased, two domains tend to appear. At slow dilution rates, the concentration in the cell phase tends toward zero; at high dilution rates, it tends toward a finite value that is not ensitive to K_S^*.

Figure 3.16 shows that this effect is due to the saturation of specific T/M rate at high affinities that intervene around a critical value D_c. For $D < D_c$, the rate rapidly reaches a value near to its maximum value (cf. (3.5.22)). Beyond this value, the specific rate "saturates" to its value, and the compound is accumulated in the cell phase.

3.5.7.3 Transition (Threshold)

The process described above can present sudden changes. However, the truly discontinuous situation only intervenes in the special case where affinity is infinity and when the T/M rate in lower than washout. In reality, in the situation where $K_S^* = 0$, the solution (3.5.20) takes the following form:

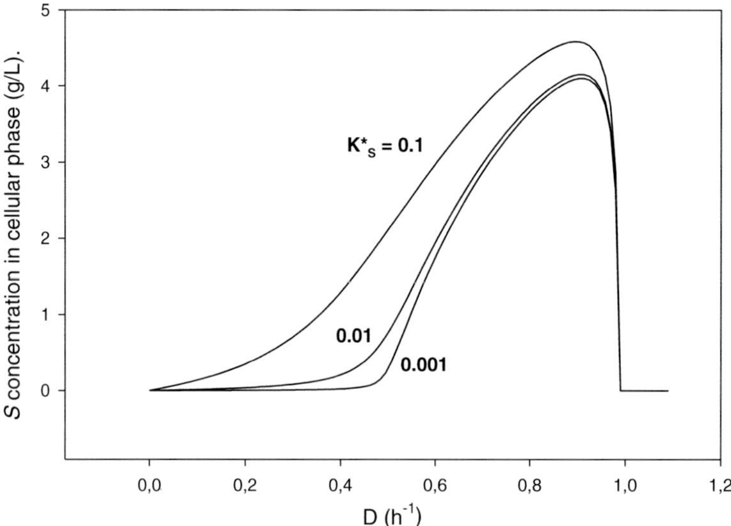

Fig. 3.15 Pseudo-homogeneous concentration variation of the compound in the cell phase according to the dilution rate. Explanations: refer to the text. Generating model (Monod) parameters: as in Fig. 3.13. PDS parameters: $V_S^0 = 0.9$ (h^{-1}); $K_S^* = 0.1$, 0.01, 0.001 (refer to curves)

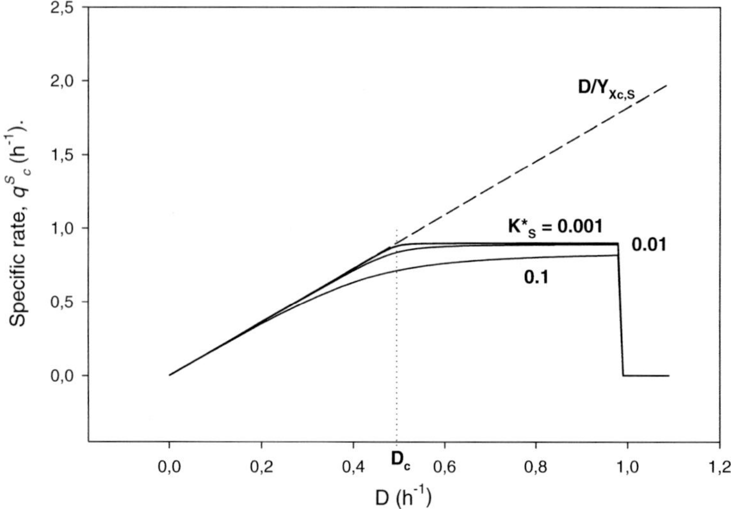

Fig. 3.16 Specific transport/metabolic rate variation according to the dilution rate. Explanations: refer to the text. Generating model (Monod) parameters: as in Fig. 3.13. PDS parameters: as in Fig. 3.15

$$\tilde{C}_S^c = \frac{-a_1 + |a_1|}{2a_2} \tag{3.5.23}$$

In these conditions, concentration in the cell phase becomes nil if $a_1 > 0$ and takes a finite value if $a_1 < 0$:

$$\tilde{C}_S^c = \frac{\Phi_S^0 - V_S^0 X^c}{D}$$

It is clear that the threshold value intervenes when $a_1 = 0$, that is to say when (refer to (3.5.19))

$$\Phi_S^0 = X^c V_S^0 \tag{3.5.24}$$

This relationship expresses the sudden change that occurs when the interphasic exchange flow becomes equal to the maximum T/M flow ($X^c V_S^0$). Using (3.5.4) and (3.5.1), it is easy to calculate D_c from (3.5.24):

$$D_c = V_S^0 Y_{X^c,S} \tag{3.5.25}$$

This equality is very important because it shows the relationship that exists between the critical dilution rate (threshold) and the global, intrinsic quantities of the culture, namely the global, yield coefficient (that is a constant equal to Y_M), and the maximum T/M rate. The relationship (3.5.25) is only exact for $K_S^* = 0$ but remains a very valid approximation for $K_S^* \ll 1$. It is evident that threshold phenomena are only observable if the critical value of dilution rate is less than the washout value.

Comments In such a system, it appears that a compound that is likely to be transported and metabolized in the cell phase can be maintained at extremely low intracellular concentrations at dilution low dilution rates and is a compound that can be suddenly accumulated when the dilution rate reaches a critical value. The conditions under which such a phenomenon occurs are,

- that it obeys a kinetic such as (3.5.16);
- that the maximum T/M rate is less than the washout;
- that affinity for the compound is sufficiently high.

If the compound is considered active (at genetic, enzymatic,… levels) the sudden physiological changes in a continuous culture can be taken into account using a representation that is as simple as this one.

3.5.7.4 n = 2—Model with Two Pathways

In order to simplify the presentation, let us consider here the situation where the system presents a pathway with a low specific rate and high affinity and a pathway with very low affinity. This situation is very often found in reality (Walker 1998; Postma et al. 1989). The pathway with high affinity has been presented in the preceding section and can be represented by (cf. (3.5.16))

$$q_S^c(h) = \frac{V_S^0 \tilde{C}_S^c}{K_S^* X^c + \tilde{C}_S^c}; \quad K_S^* \ll 1 \tag{3.5.26}$$

where h designates the kinetic with high affinity.

The other pathway is expressed in R-concentrations by the general relationship (cf. (3.5.14)):

$$q_S^c(2) = \frac{V_S^0(2) C_S^c}{K_S(2) + C_S^c} \tag{3.5.27}$$

A system where affinity is sufficiently low is chosen so that

$$K_S(2) \gg C_S^c \tag{3.5.28}$$

The relationship (3.5.27) takes the form of a kinetic of order 1,

$$q_S^c(2) \approx k_0 C_S^c \tag{3.5.29}$$

where $k_0 = V_S^0(2)/K_S(2)$.

By moving from R- to E-concentrations, as before, the expression for the specific T/M rate of the pathways with weak affinity l is obtained:

$$q_S^c(l) = k_0^* \frac{\tilde{C}_S^c}{X^c} \tag{3.5.30}$$

where $k_0^* = k_0 \delta_c$.

The total rate is then given by the sum of the two components, (3.5.26) and (3.5.30) (cf. (3.5.14)):

$$q_S^c = q_S^c(h) + q_S^c(l) = \frac{V_S^0 \tilde{C}_S^c}{K_S^* X^c + \tilde{C}_S^c} + \tilde{C}_S^c k_0^* / X^c \tag{3.5.31}$$

and the mass balance in steady state is written cf. (3.5.2) as

$$\Phi_S^0 = \frac{V_S^0 \tilde{C}_S^c}{K_S^* X^c + \tilde{C}_S^c} X^c + \tilde{C}_S^c \left(k_0^* + D\right) \tag{3.5.32}$$

The second-order polynomial with associated, variable coefficients is,

$$P^2(\tilde{C}_S^c) \equiv a_2'\left(\tilde{C}_S^c\right)^2 + a_1'\tilde{C}_S^c + a_0' = 0 \tag{3.5.33}$$

with

$$\begin{aligned} a_2' &= -K_S^* X^c \Phi_S^0 \\ a_1' &= X^c\left(V_S^0 + K_S^*\left(D + k_0^*\right)\right) - \Phi_S^0 \\ a_0' &= D + k_0^* \end{aligned} \tag{3.5.34}$$

Equation (3.5.33) possesses the same properties as (3.5.18) and admits only one real nonnegative solution,

$$\tilde{C}_s^c = \frac{-a_1' + \sqrt{a_1'^2 - 4a_2'a_0'}}{2a_2'} \tag{3.5.35}$$

Figures 3.17 and 3.18 show the essential differences between models with one and two pathways.

The simulations show that the concentration in the cell phase can be reduced by increasing k_0^*. The pathway with low affinity then makes it possible to regulate the quantity of product in the cell (compare Figs. 3.17 and 3.15 where the maximum concentration is reduced by about four times). The total specific T/M rate no longer

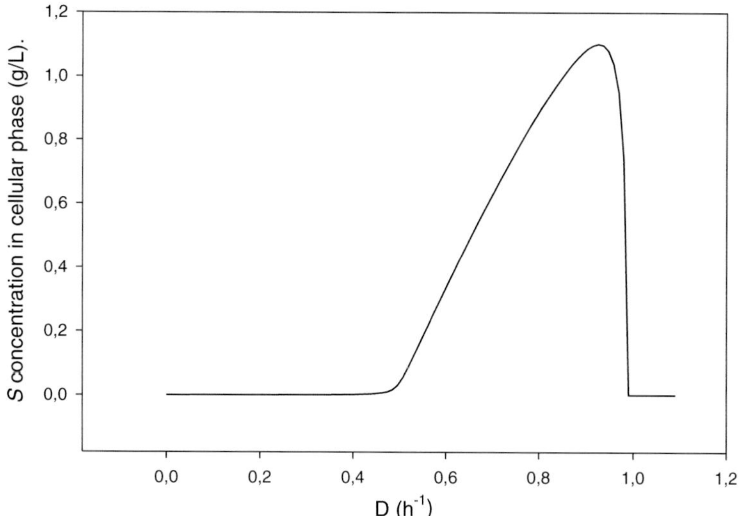

Fig. 3.17 Pseudo-homogeneous concentration variation of the compound in the cell phase according to the dilution rate for $n = 2$. Explanations: refer to the text. Generating Model (Monod) parameters: as in Fig. 3.13. PDS parameters: $V_S^0 = 0.9$ (h^{-1}); $K_S^* = 1.10^{-4}$; $k_0^* = 2.5$ (h^{-1})

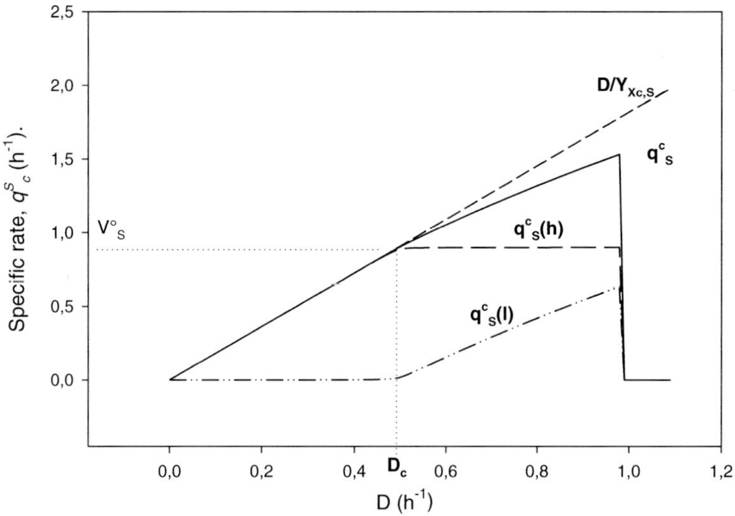

Fig. 3.18 Specific transport/metabolism rate variation according to dilution rate for n = 2. Explanations: refer to the text. Generating Model (Monod) parameters: as in Fig. 3.13. PDS parameters: as in Fig. 3.17

undergoes the phenomena of saturation. This rate and the two components (high and low) are represented in Fig. 3.18. In this example, it appears clearly that for $D < D_c$, the concentration of the compound in the cell phase is nearly nil. Consequently, the specific T/M rate is also nearly nil since the kinetic is of the order of 1. The pathway with high affinity because of its hyperbolic form, can take a significant, finite value even for low values of concentrations (providing the affinity term i of the same order of size as the concentration, as is the case at high affinities).

Beyond D_c, the rate of the high-affinity pathway saturates, as before, but the low-affinity p takes significant values, depending linearly on D. With the values for parameters used, the total specific rate is close to its maximum value for $D < D_c$ and moves away from this value when D increases. Globally, this situation is comparable to the situation represented in Fig. 3.14, when the mechanisms in play are totally different. (Note, however, that the model with one pathway is only a special case of the model with two pathways when $k_0^* = 0$.)

3.5.7.5 Transition

By following the same reasoning as above, it can be shown that the change is effected for $a'_1 = 0$. So,

$$\Phi_S^0 = X^c\left(V_S^0 + K_S^*\left(D + k_0^*\right)\right) \tag{3.5.36}$$

At high affinities the following approximation is obtained:

$$\Phi_S^0 \approx X^c\left(V_S^0 + K_S^* k_0^*\right) \tag{3.5.37}$$

The last term in this relationship only has a significant value if k_0^* is very high (of the order of the inverse of K_S^*). In these conditions, (3.5.37) signifies that the change intervenes when the interphasic exchange flux is equal to the maximum flux of the pathways with high affinity plus a "crossed term" ($X^c K_S^* k_0^*$) that shows an interaction between the two pathways. However, with practical examples, we have never noticed this situation, it is probable that the approximation

$$\Phi_S^0 \approx X^c V_S^0 \tag{3.5.38}$$

is sufficient in the majority of cases.

From the point of view of critical dilution, it is shown that at high affinities

$$D_c = \frac{V_S^0 + K_S^* k_0^*}{1/Y_{X^c,S} - K_S^*} \tag{3.5.39}$$

that comes back to

$$D_c \approx \left(V_S^0 + K_S^* k_0^*\right) Y_{X^c,S} \tag{3.5.40}$$

(case corresponding to (3.5.37)). Once again, it is probable that in many cases, the following approximation will suffice:

$$D_c \approx V_S^0\, Y_{X^c,S} \tag{3.5.41}$$

A value equal to that obtained with the model with one pathway (cf. 3.5.25). Once again, this relationship can have great practical importance when modification of the critical value is desired by changing the culture conditions (media, genetic modification, etc.).

Comments A system of which the total specific T/M rate is composed of two terms (high and low affinities) presents the feature of being able to suddenly change one metabolic way to another when a critical value (threshold) for the dilution rate is reached. In the case under consideration, the low-affinity pathway is suddenly activated when the high-affinity pathway takes a constant value (for saturation, Fig. 3.18). Above the threshold, the concentration of the compound increases in the cell phase (Fig. 3.17). This effect can, according to the nature of the compound, lead to consequences on other metabolic pathways and constitute a chemical signal at cell level. It is interesting to note that the concept introduced here differs noticeably from that considered for the intercellular communication. For example, Decho (1999), describes quorum sensing by an effector (homoserine lactone) that penetrates cells by diffusion. The effector would only be active, by linking to receptors above a critical concentration (threshold) in the extracellular medium. In the approach used, a supplementary condition appears, namely the physiological state of the cell. The signal would not be activated according to exterior concentration alone, but would also depend on the rate of metabolism of the effector in the cell. In this case, cells that are identical but which have different growth rates would not necessarily give the same response to a common effector even for identical extracellular concentrations.

3.5.7.6 Model with Two Pathways with Metabolite Release

Before approaching more complex cases, a somewhat special case must be considered where part of the compound that has entered into the cell phase leaves it again. As such, this situation has only one point of interest here, but it is a necessary stage in the reasoning.

Let us look again at the mass balance (3.5.32) in the implicit form,

$$\Phi_S^0 - \left(q_S^c(h) + q_S^c(l)\right)X^c - D\tilde{C}_S^c = 0 \tag{3.5.42}$$

and let us suppose that a part of the intracellular flux of the low-affinity pathway is excreted, where

- $\beta q_S^c(l)X^c$ is the fraction that remains in the cell phase and
- $(1-\beta)q_S^c(l)X^c$ the excreted fraction

necessarily with $0 \leq \beta \leq 1$.

With one part of the compound excreted, it would be necessary to write the mass balance in the form

$$\Phi_S^0 - \left(q_S^c(h) + \beta q_S^c(l)\right)X^c - D\tilde{C}_S^c = 0 \tag{3.5.43}$$

The simulations show that (3.5.43) gives rise to a violation of (3.5.9), and that the total specific T/M rate exceeds its maximum rate when $D > D_c$. The mass balances (3.5.42) and (3.5.43) are, however, theoretically correct but (3.5.43) violates one of the conditions on the maximum specific rate when Monod's generating model is used (cf. (3.5.21a, b)). One way to satisfy both balances simultaneously consists in finding an operator (OP) that acts on the biomass to adjust it,

$$X^c \xrightarrow{\text{OP}} X^{/c} \tag{3.5.44}$$

This operator is easy to find, since the modified biomass must also satisfy the balance (3.5.42). From this relationship, it is easily derived that

$$X^{/c} = \frac{\Phi_S^0 - D\tilde{C}_S^c}{q_S^c(h) + q_S^c(l)} \tag{3.5.45}$$

By comparing (3.5.42) with the modified biomass and (3.5.43) it is easily shown that

$$\frac{X^{/c}}{X^c} = \frac{q_S^c(h) + \beta q_S^c(l)}{q_S^c(h) + q_S^c(l)} \leq 1 \tag{3.5.46}$$

and so the adjustment of the biomass consists in a reduction of this as against the generating model.

In practice, the change (3.5.44) that consists in modifying the biomass to reestablish a mass balance that is disturbed by an outlet of the compound is a return to partial decoupling of the metabolism of this compound and production of biomass. It is clear that this process applies to particular compounds, namely the substrates. In other terms, the relationship (3.5.45) gives the quantity of biomass formed when part of the substrate is away from the biosynthesis pathways (anabolism and associated fueling).

Figure 3.19 shows the result of the change (3.5.45). Note that the critical value D_c is unchanged as is the washout value. The fall in biomass is all the more sudden as affinity rises. Figure 3.20 shows the profiles of the specific rates. Below the threshold, the total specific rate is unchanged compared with the simple model with two pathways since the low-affinity compound is inactive.

Above the threshold, the total rate suddenly increases since the interphasic flux remains unchanged, whether there is excretion or not, but the biomass reduces

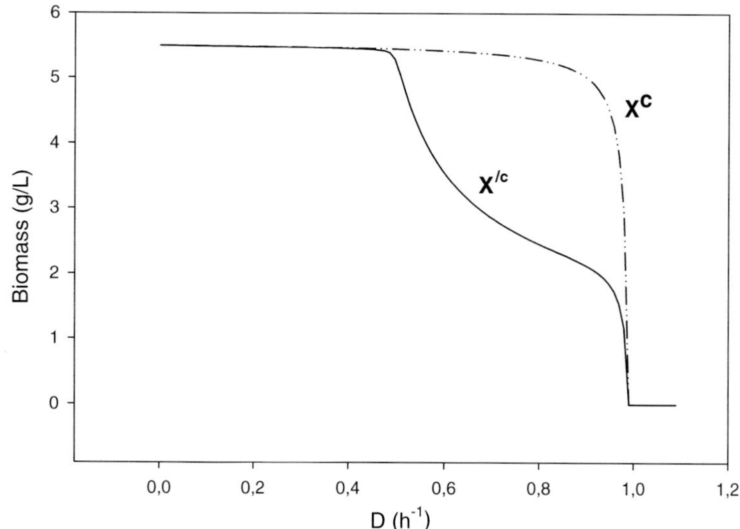

Fig. 3.19 Biomass variation according to the dilution rate. The change (3.5.45) generates a sudden fall in biomass in the steady state for $D > D_c$ (*solid line*). That of the generating model keeps its usual profile (*dotted line*). (The two critical values D_c and D_W are preserved). Generating model (Monod) parameters: as in Fig. 3.14. PDS parameters: $V_S^0 = 0.9$ (h^{-1}); $K_S^* = 1.10^{-4}$; $k_0^* = 2.5$ (h^{-1}); $\beta = 0.2$

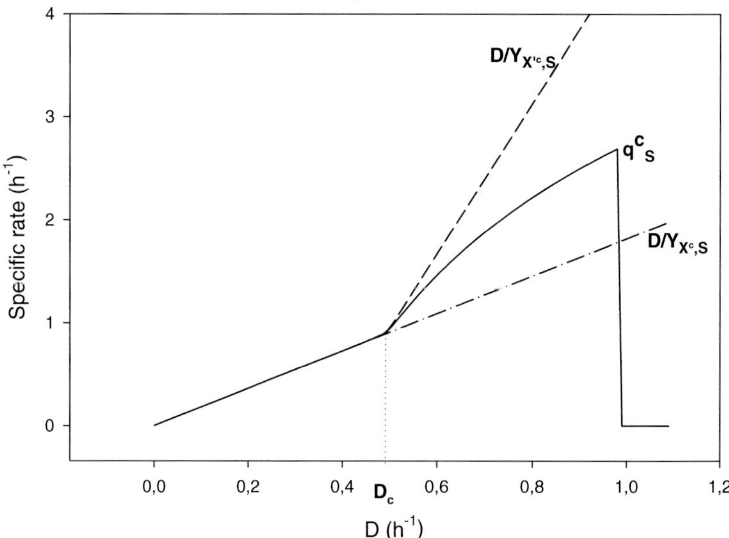

Fig. 3.20 Specific variation of transport/metabolic rate according to dilution rate. The total specific speed, as well as its maximum values before and after transition, are represented. For $D > D_c$, observe that the specific speed exceeds the maximum value before the change. The adjustment in biomass (3.5.45) gives rise to a sudden increase in the maximum value of total specific speed. Generating model (Monod's) parameters: as in Fig. 3.13. PDS parameters: as in Fig. 3.19

suddenly. This fully explains the fact that the same substrate flux is handled by a smaller quantity of biomass, so there is an increase in specific rate (flux by biomass unit).

It is important to realize that the simulation that makes it possible to obtain the result in Fig. 3.20 does not necessitate any conditional test whatsoever during the course of the program, nor any optimization condition. The algorithm is very simple and is as follows:

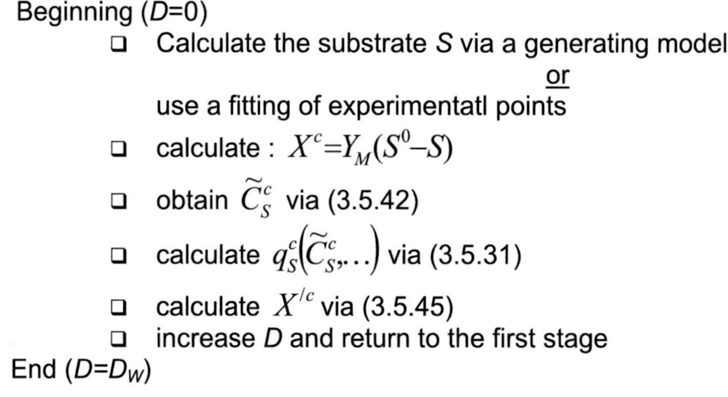

Beginning (D=0)

- ❑ Calculate the substrate S via a generating model
  or
  use a fitting of experimentatl points
- ❑ calculate : $X^c = Y_M(S^0 - S)$
- ❑ obtain $\widetilde{C}_S^c$ via (3.5.42)
- ❑ calculate $q_S^c\left(\widetilde{C}_S^c, \ldots\right)$ via (3.5.31)
- ❑ calculate $X^{/c}$ via (3.5.45)
- ❑ increase D and return to the first stage

End (D=D_W)

For the algorithm, the yield coefficient used is the one from the generating model that makes it possible to evaluate X^c. It is clear that the yield coefficient is no longer constant over the entire interval D; it is calculated using the usual relationship (3.5.4) with the value $X^{/c}$.

For the determination of D_c, the relevant value of the yield coefficient is that obtained for $D < D_c$ where in the absence of maintenance term $Y_{X^c,s} = Y_{X^{/c},s} = Y_M$.

3.5.7.7 Excretion of Metabolites

In the preceding section, an outlet of substrate from the cell phase was considered. Although this phenomena can be considered (refer to Sect. 3.4), the substrate is not generally excreted in a big way in the initial form. In particular, with phenomena of fermentation, part of the substrate is diverted from the biosynthetic flow and from fuelling, transformed by redox reactions and secreted in another form. It is this type of phenomena that is under consideration here. Before getting to the heart of the matter, it is appropriate to introduce some general notions.

General Remarks In the PDS approach, all flows and concentrations are expressed in mass form. Introduced here are mainly the relationships between mass quantities and molar quantities.

For all general chemical reactions of the type

$$\nu_A A + \nu_B B + \ldots \rightarrow \nu_P P + \nu_Q Q + \ldots \quad (3.5.47)$$

There are two elementary relationships (for the sake of simplicity, let us use the index alone instead of the symbol for the compound: $i = P_i$):

$$\sum_i \nu_i \text{MM}_i = 0 \quad (3.5.48)$$

That which expresses the molecular mass balance, with MM_z, the molar mass of compound z and ν_z its stoichiometric coefficient (defined as negative on the left and positive on the right).

Moreover, for each pair of compounds,

$$\frac{n_i}{n_j} = \left|\frac{\nu_i}{\nu_j}\right| \quad (3.5.49)$$

where n_z is the number of moles of compound z. (The molar relationship always being positive, absolute values must be used for stoichiometric coefficients.) This relationship expresses that the ratio of the number of moles is equal to the ratio of stoichiometric coefficients.

Let us consider a special case (3.5.47), now,

$$\nu_S S \rightarrow \ldots \rightarrow \nu_{1.} P_1 + \ldots \nu_j P_j + \ldots \tag{3.5.50}$$

This equation only represents the initial reaction (S) and the group of products (P_i), independently of the intermediary compounds and of the pathway followed to obtain the final compounds. Equation (3.5.48) can be put in the form

$$-\nu_S \mathrm{MM}_S = \sum_j \nu_j \mathrm{MM}_j \;\; (j \neq S) \tag{3.5.51}$$

And by normalizing to MM_S:

$$1 = -\sum_j \frac{\nu_j \mathrm{MM}_j}{\nu_S \mathrm{MM}_S} (j \neq S) \tag{3.5.52}$$

Let us define the positive value,

$$\rho_{j,S} = \left| \frac{\nu_j \mathrm{MM}_j}{\nu_S \mathrm{MM}_S} \right| \tag{3.5.53}$$

There are the following properties:

$$\sum_j \rho_{j,S} = 1 \tag{3.5.54}$$

$$\rho_{j,S} = \frac{1}{\rho_{S,j}} \tag{3.5.55}$$

Using the relationship of the number of moles to mass

$$n_k = \frac{M_k}{\mathrm{MM}_k} \tag{3.5.56}$$

(where M_k is the mass of the compound k) in the relationship (3.5.49), it is easy to show that for each product,

$$\frac{M_i}{M_S} = \left| \frac{\nu_i \mathrm{MM}_i}{\nu_S \mathrm{MM}_S} \right| \tag{3.5.57}$$

By comparing with (3.5.53), it is clear that

$$\rho_{i,S} = \frac{M_i}{M_S} \tag{3.5.58}$$

This will be called the mass ratio coefficient (MRC).

3.5.7.8 Application to Concentrations

By putting (3.5.58) in the form

$$M_i = \rho_{i,S} M_S \tag{3.5.59}$$

it is easy to show that at constant volume

$$\tilde{C}_i^x = \rho_{i,S} \tilde{C}_S^x \ ; \ x = c, m, \ldots \tag{3.5.60}$$

(This relationship obviously extends to other types of concentrations, R-concentrations, etc.)

3.5.7.9 Application to Specific Rates

The relationship (3.5.58) can be changed to the following identity:

$$\frac{\Delta M_i}{\Delta t X^{/c}} \frac{\Delta t X^{/c}}{\Delta M_S} = \rho_{i,S} \tag{3.5.61}$$

When writing the balances, the specific rates are defined as positive (and have the correct sign). So for a single-cell schema such as (3.5.50) it can be written as

$$\lim_{\Delta t \to 0} = \frac{\Delta M_k}{\Delta t X^{/c}} \equiv q_k^x; \quad x = m, c, \ldots \tag{3.5.62}$$

and so (3.5.61) takes the form

$$\frac{q_i^x}{q_S^x} = \rho_{i,S} \tag{3.5.63}$$

A special form of (3.5.63) is as follows:

$$q_i^c = \rho_{i,S} q_S^c \tag{3.5.64}$$

so the equivalent form, in our usual formalism is

$$q_{Pi}^c = \rho_{i,S} q_S^c \tag{3.5.65}$$

3.5.7.10 Mass Balances

In the cell phase, each product (or metabolite) excreted is characterized by the balance,

$$\frac{\mathrm{d}\tilde{C}^c_{Pi}}{\mathrm{d}t} = \Pi^c_{Pi} - \Phi^0_{Pi,c}(m) - D\tilde{C}^c_{Pi} + \tilde{C}^c_{Pi}\frac{\mathrm{d}\ln N^c_T}{\mathrm{d}t} \tag{3.5.66}$$

where

Π^c_{Pi} is a production flow (for example, from the substrate, as in the reaction scheme (3.5.50)).

$\Phi^0_{Pi,c}(m)$ is the net flow (inflow/outflow balance) of the product excreted from the cell phase to the matrix phase.

The other terms keep their previous significance.

In the matrix phase, if P_i does not undergo any change whatsoever ($R^m_{Pi} \equiv 0$), the balance is written as,

$$\frac{\mathrm{d}\tilde{C}^m_{Pi}}{\mathrm{d}t} = \Phi^0_{Pi,c}(m) - D\tilde{C}^m_{Pi} \tag{3.5.67}$$

The stationary state corresponding to (3.5.66) implies that

$$\Phi^0_{Pi,c}(m) = \Pi^c_{Pi} - D\tilde{C}^c_{Pi} \tag{3.5.68}$$

which expresses that the flow excreted is equal to the production rate less the hydraulic outlet term of the product associated with the cell phase. Just as (3.5.67) gives

$$\Phi^0_{Pi,c}(m) = D\tilde{C}^m_{Pi} \tag{3.5.69}$$

expressing that the excretion flux is compensated for by the hydraulic outlet in the matrix phase.

By combining (3.5.68) and (3.5.69),

$$\Pi^c_{Pi} - D\left(\tilde{C}^c_{Pi,c} + \tilde{C}^m_{Pi,c}\right) = 0 \tag{3.5.70}$$

for systems where the biomass is not too high and accumulation of the product in the cell phase is low, it is reasonable to make the hypothesis that

$$\tilde{C}^c_{Pi} \ll \tilde{C}^m_{Pi} \tag{3.5.71}$$

Note This hypothesis does not imply that the real intracellular concentration is insignificant, but only that the pseudo-homogeneous concentration in the cell phase is negligible compared to the concentration in the medium.

When (3.5.71) applies, relationship (3.5.70) gives the approximate form,

$$\Pi^c_{Pi} \approx D\tilde{C}^m_{Pi} \tag{3.5.72}$$

This relationship establishes the link between the production flow in the cell phase and concentration of metabolite n excreted in the matrix phase. From this it is deduced that

$$\tilde{C}^{m}_{Pi} = \frac{\Pi^{c}_{Pi}}{D} \tag{3.5.73}$$

3.5.7.11 Relationship with the Kinetic Expressed in Terms of the Substrate

For evaluation of the production of metabolite on the basis of kinetics expressed in terms of the substrate, the link between the equivalent substrate and the product must be established. In a general way, a flow can be represented by the product of a specific rate with biomass,

$$\Pi^{c}_{Pi} \equiv q^{c}_{Pi} X^{/c} \tag{3.5.74}$$

The link between specific production rate of the excreted metabolite and the rate of disappearance of the substrate is then given by (3.5.65). Using this relationship and combining (3.5.73) and (3.5.74), it is found that (for a diagram such as (3.5.50))

$$\tilde{C}^{m}_{Pi} = \rho_{Pi,S} q^{c}_{S}(^{*}) X^{/c}/D \tag{3.5.75}$$

It is evident that all the substrate that is treated in the cell phase will not be excreted in the form of metabolites. $q^{c}_{S}(^{*})$ is here, the fraction of the specific rate engaged in the production of metabolites P_i. The relationship (3.5.75) is general in the sense that it does not depend on a precise model (it is an implicit model). The explicit form of (3.5.75) can be obtained using the fraction of production as defined in the model with two pathways with compound release. So,

$$\tilde{C}^{m}_{Pi} = \rho_{Pi,S}(1-\beta) q^{c}_{S}(l) X^{/c}/D \tag{3.5.76}$$

This relationship makes it possible to calculate the concentration of a metabolite excreted in the matrix on the basis of the T/M kinetic of the substrate. This is obviously a fundamental result in the evaluation of the validity of the model when it is applied to a growth decoupling process and the use of the substrate. The applications concern, for example, the process of fermentation. The example of ethanol production by *Saccharomyces cerevisiae* will be considered in detail in the following section.

By way of example, Fig. 3.21 shows the profile of concentrations of excreted products. Curve (1) shows the excretion of equivalent substrate ($\rho_{P,S} = 1$) and curves (2) and (3), those for the hypothetical products, respectively, represent 3/4 and 1/4 of the mass of the substrate ($\rho_{P1,S} = 0.75$ and $\rho_{P2,S} = 0.25$). The curves qualitatively tend well toward those usually observed in a chemostat. The corresponding specific

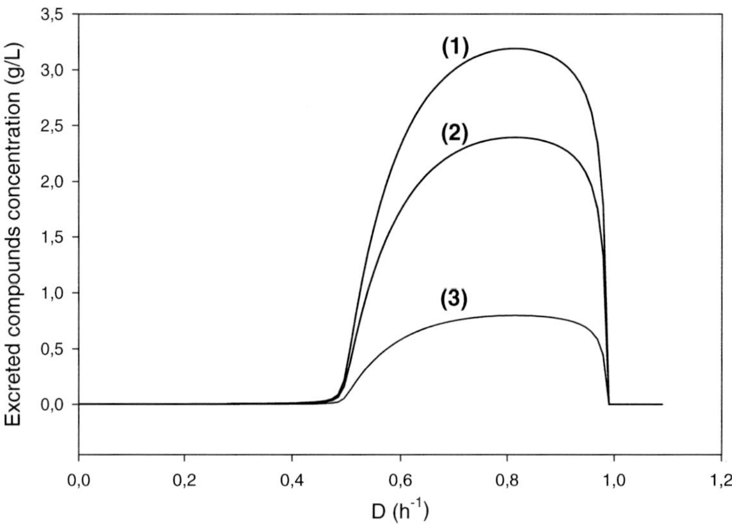

Fig. 3.21 Concentration of excreted compounds in the matrix phase according to the dilution rate. The curve (*1*) shows the equivalent substrate profile ($\rho_{P,S} = 1$). The curves (*2*) and (*3*) correspond to the diagram $S \rightarrow P_1 + P_2$ with $\rho_{P1,S} = 0.75$ and $\rho_{P2,S} = 0.25$. The substrate is degraded into two products, P_1 and P_2, with hypothetical mass ratios of 3/4 and 1/4, respectively (cf. 3.5.53 and 3.5.58)

rates (not represented) vary in a monotonously increasing manner with D above the critical threshold and are nil below it. They tend toward a straight line when k_0^* is sufficiently high.

Remark In all cases with the threshold presented beforehand, it is always possible to calculate a substrate threshold value. It is this value for which $D = D_c$ in (3.5.21a). This relationship is continuous for dilution rates that are less than washout and this substrate value has scarcely any significance. Moreover, it is rare that (3.5.21a) is the best adjustment of the experimental points. The imprecision of measurements of the residual substrate, within the range of D, makes this measurement completely nonfunctional.

3.5.8 Discussion

From a conceptual point of view, the point that deserves our attention most is the expression of the transport/metabolism kinetics in terms of pseudo-homogeneous concentrations (E-concentrations) at the beginning of a representation in terms of local concentrations (R-concentrations). In a polyphasic representation, the T/M kinetic describes a process that continues at the interface of two phases. So it is a question, in fact, of a heterogeneous kinetic.

This situation already presents, in a general manner, problems at the level of definition of the efficient concentrations (Villermaux 1982; Roels 1983). Coulson and Richardson (1987) have discussed the local form (3.5.14) for $n = 1$. They conclude that this relationship expresses well both the transport and the metabolism of the substrate, on condition that the maximum rate of metabolism is greater than the rate of transport. This condition is, however, more restrictive because it excludes the possibility of accumulation of free substrate in the cell. Moreover, the hyperbolic form of (3.5.14) reflects well the phenomena of transport by diffusion (whether facilitated or not) (Walker 1998; Schechter 1997) and is regularly used to model other transport mechanisms. The local form of the T/M phenomena expressed in terms of R-concentration can therefore be considered as adequate. On the other hand, global forms cause a dependance to appear between the local kinetic and the biomass. Without entering into details, it is emphasized that there is a similarity between the concepts proposed by Abbott and Nelsestuen (1988) and the results obtained by a completely different approach. These authors have shown that if the number of cell receptors is high enough (of the order of 10 % of the membrane surface), the ligand–receptor complex formation rate no longer depends on receptors concentration but only on cellular concentration. The authors call this phenomena the "collisional limit." Moreover, they have shown that for a hyperbolic kinetic, the collisional limit led to a variation in affinity according to the number of cells. In our model, the resulting affinity ($K_S^* X^{/c}$) is a decreasing function of the biomass if K_S^* is a true constant and if δ_c only slightly varies with culture conditions, which can be considered reasonable (Kubitschek et al. 1983, 1984; Baldwin and Kubitschek 1984). Moreover, the influence of biomass on affinity for the substrate has been defended by other authors in several experimental situations. Let us cite the growth phenomena on a simple substrate (Contois 1959) or on complex substrates (Roques et al. 1982). It is possible that this problem is complex and remains an open question. By waiting for a unified interpretation, the kinetics used here can be considered a useful phenomenological representation of the T/M kinetic, because it has been tested with success in several practical cases (among which is that in the following section). All the same, this notion is introduced in a very natural way in the PDS approach which reinforces the idea that this method of representation is adequate. Finally, it seems natural to arrive at the conclusion that sharing the substrate between a large number of cells necessitates a representation that takes into account a form of interaction between the cells. Independence of the cells that overtake on a common substrate only appear "intuitive" for systems at very weak cell concentration where the cells can be considered as isolated one from another.

Altogether, models with one or two pathways make it possible to take into account signal phenomena appearing at a critical dilution rate by invasion of the cell by a compound that is transported and partially metabolized. The sudden increase of an effector in the cell medium is likely to set off (in)activation phenomena at the level of enzymatic as well as genetic systems. The theory makes this possible without any ad hoc conditions, nor any preceding constraint whatsoever being specified, and this by using an extremely simple algorithm.

When the excretion of metabolite is considered, the approach makes it possible to model decoupling between consumption of substrate for growth and the fraction used for other ends, such as excretion of fermentation products. This last approach is particularly fruitful since it results from the possibility of visualizing the distribution of the main specific rates (or principle flows) of the metabolism of the substrate. Figure 3.22 shows one way of representing this distribution in the cell phase (represented as a simple cell).

The arrow (1) denotes the total T/M flow of substrate in the model with two pathways. The high-affinity pathway (4) is permanent and is used for anabolism and fueling (with an unquantified outlet pathway here, for water, CO_2, etc.). Arrows (2) and (3) represent the low-affinity pathway that is inactivated above the critical threshold (an outlet pathway is also associated with this mechanism). Arrow (5) represents the free substrate, adsorbed on the membrane and/or present in the cell. In this representation, there is a nodal point that represents the branching of the main anabolic pathways and free substrate maintenance pathways. The dotted lines indicate the pathways that are activated when $D > D_c$.

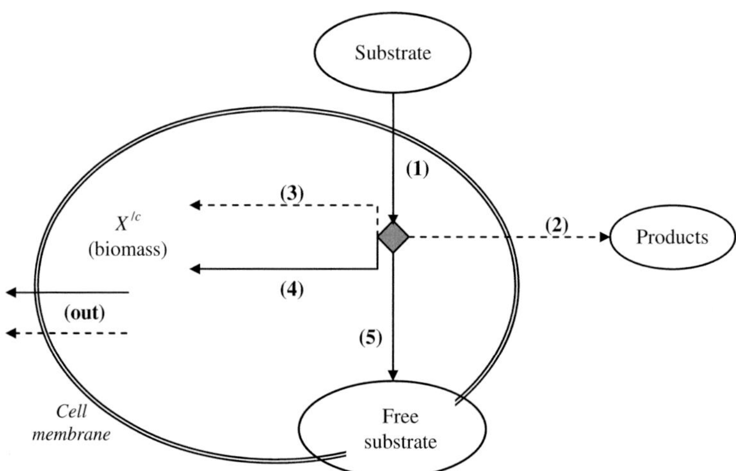

Fig. 3.22 Distribution of the principal specific rates in the cell phase (Here represented as an isolated cell). (*1*) Total T/M: $q_S^c = q_S^c(h) + q_S^c(l)$; (2) Excretion $(1 - \beta)q_S^c(l)$; Anabolism/fuelling; (*3*) $\beta q_S^c(l)$; and (*4*) $q_S^c(h)$; Free substrate $D\tilde{C}_S^c/X^{/c}$; (*out*) Outlet term taking into account the yield coefficient. (H_2O, CO_2; not quantified here). The *dotted lines* show the metabolic pathways that are activated for $D > D_c$ alone

Pathway (5) can also be nil before the critical threshold or become insignificant over the whole domain of dilution rates (this last situation depends mainly on the value of k_0^*). Globally, for $D < D_c$, only pathways (1) and (4) are active; the whole of the substrate is used for the biosynthetic reactions and fueling reactions. For $D > DC$, pathway (4) saturates and reaches its maximum value. Pathways (2), (4), and (5) are activated. Biosynthesis is then certain along pathways (3) and (4). Pathway (2) is used for the substrate transformation and for the formation of products excreted in the matrix phase.

In conclusion, it is thought that a simple model that makes it possible to take into account quite complex phenomena can have a practical use in perfecting and optimizing biotechnological procedures. From a more basic point of view, representation of the PDS has already shown its use and the perspectives that open up using explicit forms for specific rates are numerous and very encouraging.

3.5.9 Remarks on the Coupling of Maintenance Energy and the Metabolic Switch

It is perfectly possible to couple the results obtained in Sect. 3.4. concerning maintenance energy and the effects of regulation by biomass (cf. Fig. 3.19) that intervene at the time of a metabolic switch.

Figure 3.23 shows different profiles obtained by putting this coupling into action. In theory, the basic algorithm is preserved, but a recirculation flow of substrate is added to take into account the energy loss such as it is described in the last section.

Without going into details, here are some comments on the different effects:

The generating model is again the Monod's model for which the biomass profile appears in the form of a dotted line in the four figures.

The kinetic constants of the generating model are as follows: $\mu_{\max} = 1$; $K_{\mathrm{M}} = 0.1$; $Y_M = 0.55$; and the inlet concentration of the limiting substrate is $S^0 = 10$ g/L.

The kinetic constants for the PDS representation are $V_S^0 = 0.9$; $k_0 = 2.5$ and $\beta = 0.5$.

Figure 3.23a, b shows high-affinity systems with $K_S = 1 \times 10^{-4}$. This characteristic shows well-marked threshold effects (strong discontinuity). Figure 3.23a has low maintenance energy ($\Delta\Phi = +0.05$; recirculation measurement) and Fig. 3.23b shows a higher recirculation of substrate ($\Delta\Phi = +0.25$).

Figure 3.23c, d shows systems with lower affinities with $K_S = 1 \times 10^{-2}$. The threshold effects are masked and only a bump appears in the profile. Figure 3.23c has low maintenance energy ($\Delta\Phi = +0.05$) and Fig. 3.23d, a raised recirculation of substrate ($\Delta\Phi = +0.25$). The different profiles illustrate well the wealth of behaviors that the model can generate.

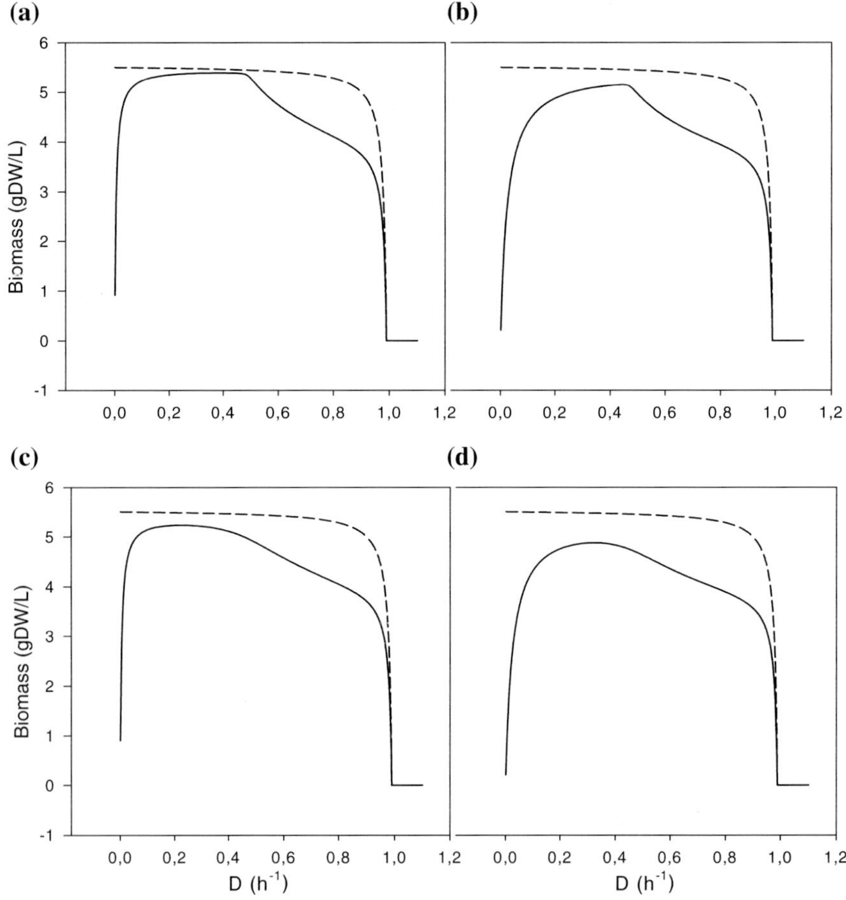

Fig. 3.23 Various steady states biomass profiles when substrate recirculation is coupled to a metabolic switch. Different profiles were obtained via a generating model (*Monod* refer to the text) and applying the two pathways model with metabolite excretion. Aspect of the profiles depend on the coupling strength. For more details, refer to the text

3.6 The Crabtree Effect in *Saccharomyces cerevisiae*

3.6.1 Introduction

It is well known that microorganisms' capacities of adaptation are considerable when faced with changes to external constraints. However, in a biological system, such as a chemostat, most observable macroscopic quantities (biomass, concentration of substrate, etc.) generally develop in a continuous manner when environmental constraints change in a continuous manner. However, there are remarkable exceptions to this usual rule and the long-term Crabtree effect is one of them. Phenomenologically, it is observed that at low dilution rates (D), yeasts grow with a constant substrate yield coefficient and keep the biomass in the chemostat relatively constant. Oxygen-specific consumption increases in a linear manner with D, and the metabolism of substrates (glucose, for example) is done by a purely respiratory (or oxidative) pathway. For a certain class of yeasts (called "Crabtree positives" or C+), there is a critical threshold for dilution rate (D_c) beyond which a net decrease in yield coefficient is observed, and a reduction in biomass. For this same value for D_c, specific consumption of oxygen becomes about constant and metabolism changes from a respiratory regime to a respiro-fermentative regime. Fermentation products subsequently appear in the medium, such as ethanol. So, for example, the biomass profile according to dilution rate presents a net discontinuity for $D = D_c$. This discontinuity is particularly well marked and precise and, according to previous experience it is impossible to work exactly to this critical value. Experimentally, experimenters find themselves either in the area "before" or the area "after" and the critical point itself seems definitely out of reach. This threshold seems to characterize a discontinuity point, in the mathematical sense of the term. The same goes for oxygen specific consumption, specific consumption of products of fermentation, etc. The whole of these observable phenomena indicates that there is a critical value for dilution rate for which the yeasts C+ show a sudden change in metabolic pathway, from respiratory to respiro-fermentative, when the hydraulic constraint on the system (dilution rate) exceeds a certain value.

The Crabtree effect has been known for a long time (Crabtree 1929) and in spite of numerous studies (Postma et al. 1989; de Deken 1966; Van Urk et al. 1989; Barford and Hall 1979; Barford et al. 1981; Cortassa and Aon 1998; Wojtczak 1996; Rieger et al. 1983; von Meyenburg 1969; Sonnleitner and Käpelli 1986; Bellgardt 2000a), a clear and unequivocal explanation has never been given till date (Walker 1998; Bellgardt 2000a). The interpretation of the phenomenon is, however, of paramount importance, notably from an industrial point of view, since those who handle yeasts tend to avoid the by-products of fermentation, having as their goal the production of biomass alone; whereas the fermentation industries have an interest in producing by-products in optimum amount and so reduce production of biomass as much as possible. Moreover, certain cancerous cells also show a Crabtree effect (for references, refer to Wojtczak 1996) and the elucidation of this can have important consequences in the medical field.

It is noted that the main general principles (existence of a threshold, regulation by biomass, and excretion of metabolites (as ethanol)) correspond to the properties of the model with two transport/metabolism (T/M) pathways with excretion, described in Sect. 3.5. Details of the mathematical developments have already been given. Nevertheless, as it was wished to give relatively independent laws in

Sect. 3.5 (very theoretical) and in Sect. 3.6 (one application) this part begins with a rapid reminder of the main results. In a biphasic chemostat, the mass balance of substrate S in the cell phase is

$$\frac{\mathrm{d}\tilde{C}_S^c}{\mathrm{d}t} = -D\tilde{C}_S^c + \Phi_{S,m}^0(c) - q_S^c(.)X^c + \tilde{C}_S^c \frac{\mathrm{d}\ln N_T^c}{\mathrm{d}t} \tag{3.6.1}$$

where D is the dilution rate; $\tilde{C}_S^c$ is the pseudo-homogeneous concentration (or E-concentration) of the compound in the cell phase; $\Phi_{S,m}^0(c)$ is the global, interphasic exchange flow (in the direction $m \to c$) per unit of volume; $q_S^c(.)$ is the specific transport/metabolism (T/M) rate of the compound in the cell phase; X^c is the biomass; and N_T^c is the total number of cells.

In the matrix phase, the mass balance of the substrate is

$$\frac{\mathrm{d}\tilde{C}_S^m}{\mathrm{d}t} = D\left(\tilde{C}_S^{m,E} - \tilde{C}_S^m\right) - \Phi_{S,m}^0(c) \tag{3.6.2}$$

where $\tilde{C}_S^{m,E}$ and $\tilde{C}_S^m$ are the concentrations of compound in the matrix phase at the inlet of the body of the reactor, respectively.

The steady states of (3.6.1) and (3.6.2) are, in order,

$$\Phi_{S,m}^0(c) = q_S^c(.)X^c + D\tilde{C}_S^c \tag{3.6.3}$$

and

$$\Phi_{S,m}^0(c) = D\left(\tilde{C}_S^{m,E} - \tilde{C}_S^m\right) \tag{3.6.4}$$

by dividing this last expression by the biomass, an expression for the specific interphasic transport rate is found (a specific rate being the flow per unit of biomass)

$$\frac{\Phi_{S,m}^0(c)}{X^c} = \frac{D}{Y_{X^c,S}} \tag{3.6.5}$$

where $Y_{X^c,S}$ is the yield coefficient ratio with the substrate:

$$Y_{X^c,S} = \frac{X^c}{\left(\tilde{C}_S^{m,E} - \tilde{C}_S^m\right)} \tag{3.6.6}$$

The relationships (3.6.1)–(3.6.4) constitute the implicit form for PDS representation in the sense that the net specific rate, $q_S^c(.)$ is not expressed as a function of state variables and parameters. Classically, this specific rate is expressed in terms of local concentrations that are relevant in the cell phase. So if it is chosen to represent $q_S^c(.)$ by a series of hyperbolic functions (refer to (3.5.14)),

$$q_S^c(.) = \sum_{i=1}^{n} \frac{V_S^0(i) C_S^c}{K_S(i) + C_S^c} \tag{3.6.7}$$

where $V_S^0(i)$ is the maximum specific rate of pathway i, K_S the inverse of infinity for the substrate in pathway i, C_S^c is the local concentration (or R-concentration) of the substrate, that is to say the intracellular concentration. The intracellular concentrations are, from an experimental point of view, quite difficult to measure and so difficult to use in a model. So if account is taken of the relationship between R- and E-concentrations (refer to (2.3.17))

$$C_S^c = \tilde{C}_S^c \frac{\delta_c}{X^c} \tag{3.6.8}$$

The local kinetics can be expressed in terms of global state variables. δ_c is the density of the cell phase in g/L, for example.

The model explains that the Carbtree effect is represented in the steady state for *S. cerevisiae* and this responds to the following criteria:

- The explicit form of the specific transport/metabolism rate is of the form (3.6.7) with $n = 2$.
- The change (3.6.8) applies.
- There are two T/M pathways ($n = 2$), one at a low maximum specific rate (less than washout) and at high affinity for the substrate and the other at very low affinity for the substrate, corresponding to a reaction of the order of one in ratio with S.

In these conditions, the net specific T/M rate becomes (for more details, refer to the preceding section):

$$q_S^c(.) = q_S^c(h) + q_S^c(l) = \frac{V_S^0 \tilde{C}_S^c}{K_S^* X^c + \tilde{C}_S^c} + \tilde{C}_S^c k_0^* / X^c \tag{3.6.9}$$

with

$$K_S^* = K_S(1)/\delta_c \tag{3.6.10}$$

$$k_0^* = k_0 \delta_c = \left(V_S^0(2)/K_S(2)\right)\delta_c \tag{3.6.11}$$

and

$$q_S^c(h) = \frac{V_S^0 \tilde{C}_S^c}{K_S^* X^c + \tilde{C}_S^c}; \quad K_S^* \ll 1 \tag{3.6.12}$$

$$q_S^c(l) = k_0^* \frac{\tilde{C}_S^c}{X^c} \tag{3.6.13}$$

Using (3.6.9) in the expression of the stationary state (3.6.3), it can be shown that the mass balance takes the form of a second degree of $\tilde{C}_S^c$ with variable coefficients,

$$P^2(\tilde{C}_S^c) \equiv a_2'\left(\tilde{C}_S^c\right)^2 + a_1'\tilde{C}_S^c + a_0' = 0 \tag{3.6.14}$$

with (by noting $\Phi_S^0 \equiv \Phi_{S,m}^0(c)$)

$$\begin{aligned} a_2' &= -K_S^* X^c \Phi_S^0 \\ a_1' &= X^c\left(V_S^0 + K_S^*(D + k_0^*)\right) - \Phi_S^0 \\ a_0' &= D + k_0^* \end{aligned} \tag{3.6.15}$$

that allows a unique, real, nonnegative solution,

$$\tilde{C}_s^c = \frac{-a_1' + \sqrt{(a_1')^2 - 4a_2'a_0'}}{2a_2'} \tag{3.6.16}$$

When the affinity of pathway 1 tends toward infinity ($K_S^* = 0$), the relationship (3.6.16) tends toward the function

$$\tilde{C}_S^c = \frac{-a_1' + |a_1'|}{2a_2'} \tag{3.6.17}$$

This function is canceled when a_1' is positive and takes a finite, positive value when a_1' is negative. It presents a sudden change from a nil value and a positive value for $a_1' = 0$. It is easy to show that the critical dilution rate (or threshold) is given by

$$D_c \approx V_S^0 Y_M \tag{3.6.18}$$

provided that K_S^* is quite small. (The signification of Y_M will be given later.) The relationship (3.6.18) is fundamental for the understanding of the phenomena that appear in the Crabtree positive yeasts. First of all the Crabtree effect can only be observed if the critical dilution rate is less than the washout rate ($D_c < D_W$); next, at Y_M constant, it is noted that the appearance of the change depends entirely on the T/M rate at high affinity. This result confirms those obtained by Van Urk et al.

(1989). From the point of view taken, independently of the mechanisms set working for transport, the Crabtree positive yeasts are those for which the maximum specific rate of the high-affinity pathway (V_S^0) is such that $D_c < D_W$; the Crabtree negative yeasts have a V_S^0 that is sufficiently raised for $D_c > D_W$ (roughly speaking, the Crabtree negative yeasts are characterized by $V_S^0 > \mu_{\max}/Y_M$).

Let us rewrite the implicit balance (3.6.3) by taking into account (3.6.9):

$$\Phi_S^0 - \left(q_S^c(h) + q_S^c(l)\right)X^c - D\tilde{C}_S^c = 0 \tag{3.6.19}$$

and let us imagine that one part of the substrate leaves the cell phase in the form of metabolites $\{P_i\}$.

Let us define

- $\beta q_S^c(l)X^c$ the fraction that remains in the cell phase (3.6.20)

- $(1-\beta)q_S^c(l)X^c$ the fraction that leaves it (3.6.21)

After excretion, the substrate in the cell balance becomes

$$\Phi_S^0 - \left(q_S^c(h) + \beta q_S^c(l)\right)X^c - D\tilde{C}_S^c = 0 \tag{3.6.22}$$

It is clear that the two balances (3.6.19) and (3.6.22) cannot be satisfied simultaneously. The solution consists in modifying the biomass in (3.6.19) in such a way as to satisfy the two balances. When (3.6.22) is satisfied, the new value of biomass is easily derived from (3.6.19):

$$X^{/c} = \frac{\Phi_S^0 - D\tilde{C}_S^c}{q_S^c(h) + q_S^c(l)} \tag{3.6.23}$$

and it can be shown that $X^{/c} \le X^c$.

Now let us describe the algorithm that makes it possible to calculate this fall in biomass. For solution (3.6.14) it is necessary to know the state variables X^c and S. The latter can be produced by an exterior model that shall be called the generating model or obtained experimentally. The generating model used previously was Monod's model. (The value of Y_M that appears in (3.6.18) is nothing other than the yield coefficient from Monod's model. Generally, it does not represent the system's true yield coefficient that is not a constant for all of D.)

The algorithm is very simple and does not require any conditional test whatsoever, nor any optimization condition as follows:

Beginning (D=0)

a. calculating $S(D)$ via a generating model
   or
   using an adjustment of experimental points
b. calculating $X^c = Y_M(S^0 - S)$
c. calculating $\tilde{C}_S^c$ via (3.6.16)
d. calculating q_S^c via (3.6.9)
e. calculating X^{lc} via (3.6.23)
f. increasing D and returning to a.

End ($D = D_W$)

(3.6.24)

Note that the computation of X^c at stage b is always possible, even without generating a model. Y_M then becomes a simple constant without any particular signification (even if it keeps the appearance of a yield).

A calculation should now be made of the metabolites excreted that are associated with the respiro-fermentative pathway (such as ethanol (EtOH) or CO_2). One of the advantages of this method is that it makes it possible to calculate specific rates in an isolated reaction diagram without necessarily knowing all the biochemical details. For example, from a substrate, representation of the intermediary stages that lead to final products $\{P_i\}$ can be avoided. However, all expressions used until now are expressed in equivalent—substrate and the concentrations (for example) should be found in "metabolites." (EtOH, CO_2, etc.). For this a mass ratio coefficient (MRC) is used that can be defined in the following manner.

Let there be a unique reaction scheme within the cell phase,

$$v_S S \longrightarrow \ldots \longrightarrow v_1 P_1 + v_2 P_2 + \ldots v_i P_i + \ldots \tag{3.6.25}$$

where υ_k are the stoichiometric coefficients (negative on the left and positive on the right). $\{P_i\}$ are the final products of the change.

The positively defined quantity is called the mass ratio coefficient:

$$\rho_{Pi,S} = \left| \frac{v_{Pi} \mathrm{MM}_{Pi}}{v_S \mathrm{MM}_S} \right| \tag{3.6.26}$$

where MM_k is the molar mass of k.

It can be shown that (refer to (3.5.60))

$$\tilde{C}^{x}_{Pi} = \rho_{Pi,S}\tilde{C}^{x}_{S}\,;\ x = c, m, \ldots \tag{3.6.27}$$

and

$$q^{c}_{Pi} = \rho_{Pi,S}q^{c}_{S} \tag{3.6.28}$$

Although the reasoning is not exactly thorough, the metabolite concentration in the matrix phase can be calculated in the following manner.

If account is taken of (3.6.21), the mass balance of the excreted substrate in the matrix phase can be put in the form (as equivalent substrate),

$$\frac{d\tilde{C}^{m}_{S}}{dt} = (1-\beta)q^{c}_{S}(l)X^{/c} - D\tilde{C}^{m}_{S} \tag{3.6.29}$$

with the steady state,

$$\tilde{C}^{m}_{S} = (1-\beta)q^{c}_{S}(l)X^{/c}/D \tag{3.6.30}$$

By multiplying the right and left sides of equation using (3.6.26) and (3.6.27), this is arrived at,

$$\tilde{C}^{m}_{Pi} = \rho_{Pi,S}(1-\beta)q^{c}_{S}(l)X^{/c}/D \tag{3.6.31}$$

This expresses the concentration of the metabolite P_i in the matrix phase. The general implicit form of (3.6.31) is

$$\tilde{C}^{m}_{Pi} = \rho_{Pi,S}q^{c}_{S}(*)X^{/c}/D \tag{3.6.32}$$

where $q^{c}_{S}(*)$ represents the fraction of the specific excreted rate expressed as equivalent substrate. The algorithm (3.6.24) and the relationship (3.6.31) are enough to calculate the fall in biomass and the concentration of metabolites associated with the excretion of the substrate.

3.6.2 Result

3.6.2.1 Preliminary Remarks

Experimental Data

Since the original article (Crabtree 1929), numerous authors have described the Crabtree effect in yeasts. We have chosen experimental data from Rieger et al.

(1983), for *S. cerevisiae*, mainly because they are formed from three series of measurements done in the same experimental conditions, with the exception of glucose concentration at the inlet to the chemostat (S^0 = 5, 10, and 30 g/L). This data makes it possible to compare the model in three different situations with the same real situation. (Data concerning glucose–ethanol mixes presented by the same authors will not be studied here. These experiments can be handled by our model but have not been developed.)

In spite of these advantages, the experimental data from Rieger et al. present notable drawbacks. First of all a distinction should be made between true experimental data (that is measured) from calculated or estimated data. External data belongs in the first category, such as dilution rate and concentration of glucose at the inlet. No measurement error is associated with these two values. If it can generally be found that the concentration at the inlet is very precise (negligible error) it usually does not mean that the same thing goes for dilution rate that is affected by a significant error. The three other measured quantities (biomass, substrate, and ethanol) are average values over at least three measurements and the standard deviation is given. These measurements generally present a very high dispersion, mainly for the measurement of glucose (for which the relative errors vary from 25 to 48 %). This situation is particularly unfavorable for parametric estimation (refer below). The precise measurement of the residual substrate is known to be often quite problematic (Roels 1983). In certain cases, the dispersion of results concerning biomass (in dry weight) is also much greater. By way of example, at S^0 = 30 g/L, the relative error over biomass can vary from only 1–30 %.

Moreover, the two other series of experimental data (the specific rates of oxygen consumption and carbon dioxide production) are not directly measured values, but calculated from approximations based partially on the fact that only CO_2 and ethanol (EtOH) are produced during fermentation. The confirmation of the homofermentative character of the culture is done by the authors on the basis of the carbon balance. Although these values are calculated values, they will be shown as experimental values from now on.

Another annoying pitfall for the use of data concerns the operating mode of the measurement of biomass. The samples are centrifuged, washed, and dried to constant weight. This method is perfectly justified, but the washing operation possibly eliminates glucose adsorbed on the cells. The quantity of glucose adsorbed is a piece of data that could be valuable for the testing of the model that integrates this information in the mass balances. Drying without precautions moreover, eliminates volatile products and makes it impossible to determine the EtOH content in the cellular phase. Nevertheless, in the probable absence of substantial accumulation of ethanol (Damari et al. 1988; Guajardo and Lagunas 1984), it has been estimated that this deprivation of information was negligible.

Parameter Estimation

Although the number of parameters to be estimated is low (only 5), the general procedure for estimation is delicate. The most simple cases (estimation of one or several parameters using two state variables ($\lambda(i) = f(X, Y)$), can be easily resolved using simple commercial softwares. However, other situation necessitates a parametric estimation that requires more than two state variables ($\lambda(i) = f(X, Y, \ldots, Z)$). In the more complex cases, a manual adjustment procedure was used that was based on a least squares method. This process provides relatively approximate results and certainly deserves the development of a more sophisticated protocol. Nevertheless, the obtained results are the same if preciseness and parametric estimation statistics can be improved. It would take too long to explain here with precision how the value of each parameter has been obtained but the main guidelines are as follows:

- A pre-estimation of Y_M is done by a simple arithmetical average of the experimental data taken before the fall in biomass ($D < D_c$). This parameter is a constant from the generating model and only corresponds occasionally with the yield coefficient of the system.
- The value for D_c is roughly estimated by a graphic method on the curve $X^c(D)$;
- A pre-estimation of V^0_{GLU} is obtained via $V^0_S \approx Y_M/D_c$ (cf. (3.6.18));
- From these pre-estimations, the theoretical curves are adjusted across the data for biomass, ethanol, and specific interphasic exchange rate (SIER). The five parameters are adjusted manually on the basis of a calculation of least squares (sum of the least squares over three curves).

The results obtained are recalled in Table 3.7.

The values in Table 3.7 will be used for all simulations.

Table 3.7 Parameter values

S^0	5	10	30	g/L
Y_M	0.480	0.474	0.474	g/g
K^*_{GLU}	1×10^{-6}	1×10^{-6}	1×10^{-6}	–
V^0_S	0.61	0.62	0.62	h^{-1}
k^*_0	25	25	25	h^{-1}
β	0.100	0.105	0.105	–
K_{GLU}[a]	1×10^{-3}	1×10^{-3}	1×10^{-3}	g/L
k_0[a]	2.5×10^{-2}	2.5×10^{-2}	2.5×10^{-2}	L/(g h)

[a]Values calculated on the basis of density of 1000 g/L using Eqs. (3.6.10) and (3.6.11)

Comments

(a) Results from the parametric estimation indicate that the values obtained really are constants, independent of the glucose concentration at the inlet of the chemostat.
(b) The estimated value of K^*_{GLU} depends greatly on the more or less acute shape of the change to the threshold ($D = D_c$). There is only one experimental point in this area, which causes great uncertainty about this parameter. It can be considered that it is only estimated to an order of size close to (between 1×10^{-5} and 1×10^{-7}; $\mathrm{p}K^*_{\mathrm{GLU}} = 6 \pm 1$). For the general representation of the phenomena, the quantities are quite insensitive to K^*_{GLU}, except exactly what concerns the steep profile of the change. In the literature, the value of the demi-saturation constant (affinity) for transport is of the order of $K_m = 1$ mM (Walsh et al. 1994), which corresponds to about 0.18 gGLU/L. A value of K_{GLU} of between 10^{-2} and 10^{-4}, for a biomass of 10 g/L corresponds to an affinity within the interval of 10^{-1} and 10^{-3} g/L. Although the lower limit of this interval is compatible with K_m, it is clear that the central value (average) is much lower than the estimation of transport affinity. In reality, the parameters do not have the same signification. The value of K_m is measured in experimental conditions in such a way that transport alone is taken into account (very rapid measurements, work at low temperature, etc.) where K_{GLU} measures an affinity resulting from transport and metabolism in a steady state. The same remark obviously goes for the maximum transport rate $V_{\max}$ and T/M rate, V_S^0.
(c) The couple $\{k_0^*, \beta\}$ is strongly linked, especially for low values of k_0^*. The precise value of this parameter can only be determined on condition that the free glucose (adsorbed or otherwise) in the cell phase is known. As noted, this experimental data is not available and a reasonable value has been chosen by estimating that the free glucose in the cell phase was low (several percent of the concentration at the inlet; Walsh et al. 1994).
(d) The value Y_M, used for the algorithm (3.6.24), is a constant within the interval D. It only coincides numerically with the yield coefficient of dilution rates that are less than the threshold. The true yield coefficient $Y_{X/c,\mathrm{GLU}}$ is obviously variable with D beyond the threshold value.

3.6.2.2 Checking the Model

Representation of the Residual Substrate

The algorithm (3.6.24) necessitates the calculation of substrate over the whole interval of dilution rates, in order to calculate the state variables at different points than those from experiment. In order to do this, two different methods were used,

Table 3.8 Substrate adjustment (EXP)

S^0 (g/L)	Function	Parameters	r^2
5	$\exp\left(a + b\frac{\ln D}{D^2}\right)$	$a = 4.217129316$ $b = 0.403164303$	1.000
10	$\exp\left(a + b\frac{\ln D}{D^2}\right)$	$a = 2.266763740$ $b = 0.297104130$	0.994
30	$a\exp\left(\frac{t1+t2}{t2}\right)^{t2}; t1 = (D-b)/c; t2 = d-1$	$a = 6.721420183$ $b = 0.574567450$ $c = 0.147947931$ $d = 2.982587130$	1.000

(i) the first, called experimental (EXP) consists in using a function of interpolation of experimental points. This interpolation function (fitting) does not have a physical meaning, but must represent a correlation coefficient. Its field of application is theoretically valid from the first to the last experimental value. However, this function is used here over an interval that is slightly extended. For these supplementary values it can be expected that (in the absence of discontinuity) the differences between the values calculated and the real values are less than the experimental average error (it is to be recalled that this is quite high). For example, Table 3.8 summarizes the function used to calculate the residual substrate over the interval $D \in [0.20, 0.41]$.

(ii) the second method uses a generating model (GM) of the Monod type (as in the previous section). In the stationary state, the residual substrate in the Monod model is

$$S = \frac{D.K_M}{\mu_{max} - D} \tag{3.6.33}$$

where μ_{max} is the maximum specific growth rate and K_M the demi-saturation constant. With the data from Rieger et al. (1983), the simultaneous estimation of these two parameters is delicate and all of these methods do not converge. So we proceeded in two stages. First, an estimation without the constraint of (3.6.33) has been done. The results are shown in Table 3.9.

It is clear that μ_{max} is a constant, whereas K_M is a function of the concentration at the inlet. The incertitude over K_M is high and low for μ_{max}. So this value is fixed at a mean value of 0.44 and readjusted (3.6.33) with this value as a constraint. The new values for K_M are recalled in Table 3.10.

Table 3.9 Adjustment of (3.6.33) without constraint

S^0 (g/L)	μ_{max} (h^{-1})	K_M (g/L)	r^2
5	0.44 ± 0.01	0.23 ± 0.08	0.95
10	0.45 ± 0.02	0.36 ± 0.16	0.91
30	0.43 ± 0.01	0.41 ± 0.15	0.96

Table 3.10 Adjustment (3.6.33) with μ_{max} = 0.44

S^0 (g/L)	K_M (g/L)	r^2
5	0.222 ± 0.018	0.95
10	0.302 ± 0.033	0.92
30	0.568 ± 0.048	0.96

These are the latest values for K_M that are more accurate and which have been used for the simulations.

Note Variation in K_M with concentration at the inlet is approximately linear. This clearly indicates that, contrary to maximum growth rate, affinity for the substrate is not a constant. Even if the fittings are valid for $r^2 > 0.95$, the culture is not well represented by Monod's model where K_M is a true constant. This confirms the approach that maintains that affinity depends not on the concentration at the inlet but on biomass (which in its turn, depends on the effect of S^0).

Biomass Representation

Once S and Y_M are fixed, representation of the biomasses via the generating model, X^c and the real system $X^{/c}$, are easily calculated as described in the algorithm (3.6.24).

Ethanol Representation

Before it is possible to continue with the parameter estimation, concentration of EtOH should be represented. One of the advantages of our method is that it makes it possible to calculate a specific rate without having to know the details of these metabolic reactions. So we can simply admit that the ethanol production is achieved via the following global reaction:

$$\mathrm{GLU} \rightarrow 2\mathrm{EtOH} + 2\mathrm{CO_2}(f) \tag{3.6.34}$$

(GLU: glucose). The symbol f indicates that it is a question of CO_2 produced by fermentation (and not by respiration). Using the mass ratio coefficient (3.6.26), it is possible to express the relationship between specific rate of EtOH and that expressed in equivalent substrate (cf. (3.6.28)):

$$q^c_{\mathrm{EtOH}} = \rho_{\mathrm{EtOH,GLU}} q^c_{\mathrm{GLU}}(*) \tag{3.6.35}$$

The fraction of EtOH is then given by the explicit form (cf. (3.6.21))

$$q^c_{\mathrm{EtOH}} = \rho_{\mathrm{ETOH,GLU}}(1-\beta) q^c_{\mathrm{GLU}}(l) \tag{3.6.36}$$

The concentration of ethanol in the matrix phase is then given by (cf. (3.6.32))

$$\tilde{C}^m_{\mathrm{EtOH}} = \frac{q^c_{\mathrm{EtOH}} X^{/c}}{D} \tag{3.6.37}$$

The same reasoning applies to CO_2 from the fermentation pathway and

$$q^c_{\mathrm{CO2}(f)} = \rho_{\mathrm{CO2}(f),\mathrm{GLU}}(1-\beta) q^c_{\mathrm{GLU}}(l) \tag{3.6.38}$$

$$\tilde{C}^m_{CO2(f)} = \frac{q^c_{CO2(f)} X^{/c}}{D} \tag{3.6.39}$$

This last pseudo-homogeneous concentration represents the CO_2 dissolved in the matrix phase, without degassing or supplementary reaction. Doubtless, this is not a quantity that is directly measurable at all pHs. The mass coefficients are calculated using the diagram (3.6.34) and the relationship (3.6.26). These values are found,

$$\rho_{\mathrm{EtOH,GLU}} = 0.511 \tag{3.6.40a}$$

and

$$\rho_{\mathrm{CO_2,GLU}} = 0.489 \tag{3.6.40b}$$

Specific Interphasic Exchange Rate (SIER)

An important derived quantity for the parametric estimation is the specific interphasic exchange rate that is defined by,

$$\mathrm{SIER} = \frac{\Phi^0_{\mathrm{GLU}}}{X^{/c}} \tag{3.6.41}$$

That is easily calculated from the experimental data.

Figures 3.24, 3.25, and 3.26 show the biomass, ethanol, and substrate profiles for the three concentrations at the inlet using the parameters from Table 3.7. The curves drawn with solid lines are obtained for a representation of the substrate by fitting (EXP), the curves in dotted lines using the generating model (GM). Figure 3.27 show the specific exchange rate (3.6.4) 1 and Fig. 3.28. The real yield coefficient is (3.6.6). Note that these two series of curves are invariable comparing to substrate concentrations at the inlet.

In a general manner, the agreement between the model and the experimental date is very good. However, it is noted that the profiles vary significantly according to the method of adopted representation of the substrate EXP or GM. In all, the biomass profile is always better in the EXP procedure; obviously, the same goes for

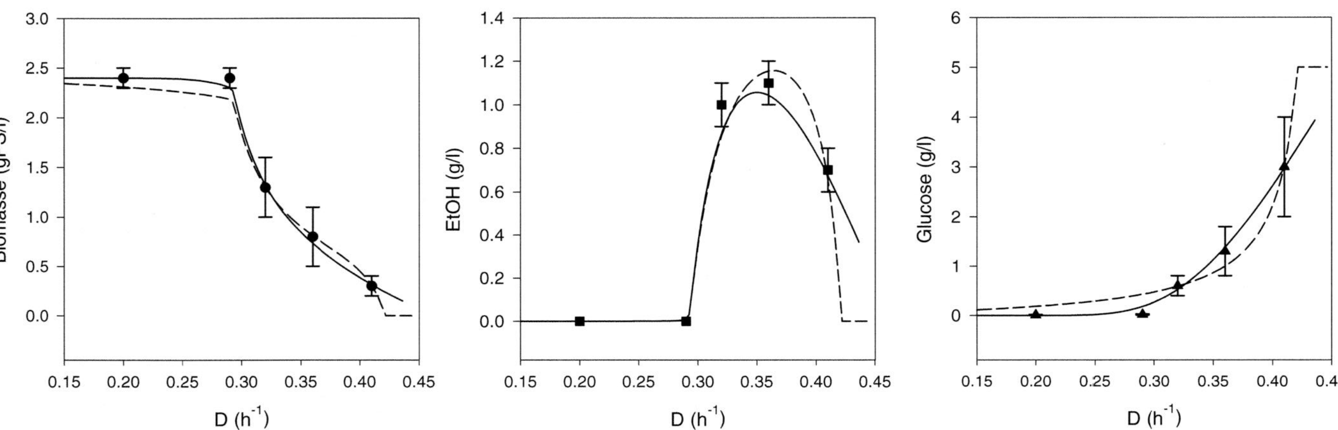

Fig. 3.24 Profiles of the principal state variables for $S^0 = 5$ g/l. *Method* The simulations are done using the algorithm (3.6.24) with values for the parameters shown in Table 3.7. The *solid lines* represent simulations done by EXP method; the *interrupted lines* are obtained by Monod's generating model (GM) method. *Comments* Note a significant difference between the substrate profiles for which one or the other model is used. Variation in ethanol profiles is more sensitive to the manner in which the substrate is represented (EXP or MG) and the variation in biomass (After Thierie (2004a, b); permission of Elsevier.)

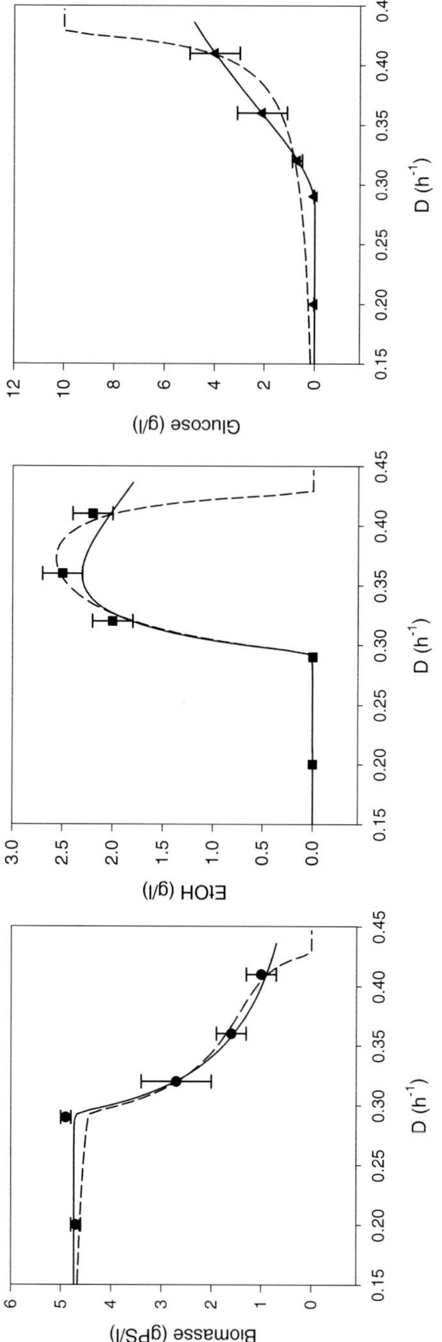

Fig. 3.25 Profiles of the principal state variables for S^0 = 10 g/l. *Method and Comments* as in Fig. 3.24 (After Thierie (2004a,b); permission of Elsevier.)

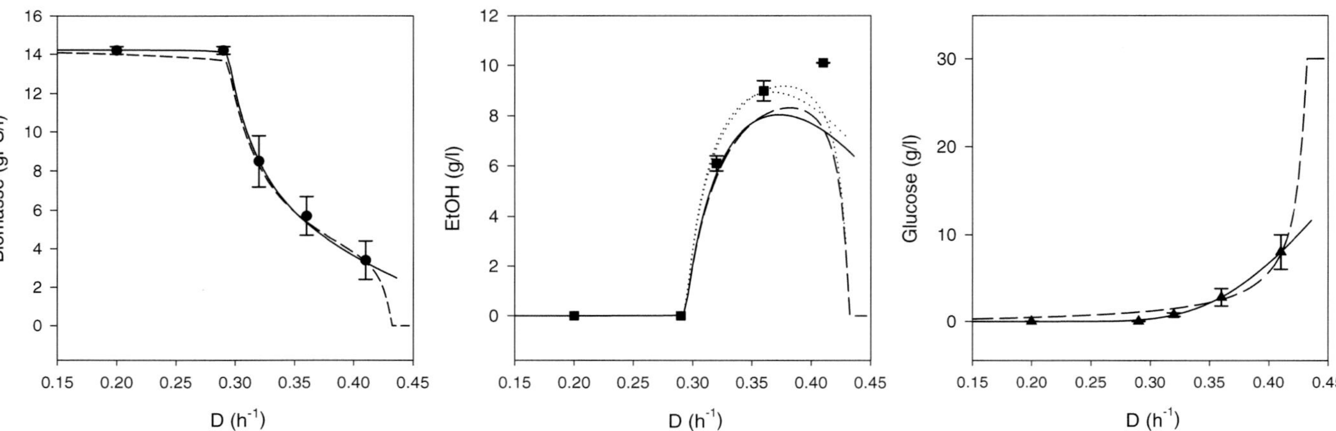

Fig. 3.26 Profiles of the principal state variables for S^0 = 30 g/l. *Method* as in Fig. 3.24. *Comments* Note a better convergence between the EXP method and the GM method for the representation of the substrate, with consequent very close profiles of both biomass and ethanol. Note, however, less good agreement between experimental data for ethanol and the simulations. The *dotted lines* represent simulations with the correcting term ε = 1.034 (refer to the relationship (3.6.42)). Even with this correction (slight), the last experimental point remains significantly distant from the calculated profiles. This experimental value is thought doubtful due to its singularity (for discussion, refer to text) (After Thierie (2004a,b); permission of Elsevier.)

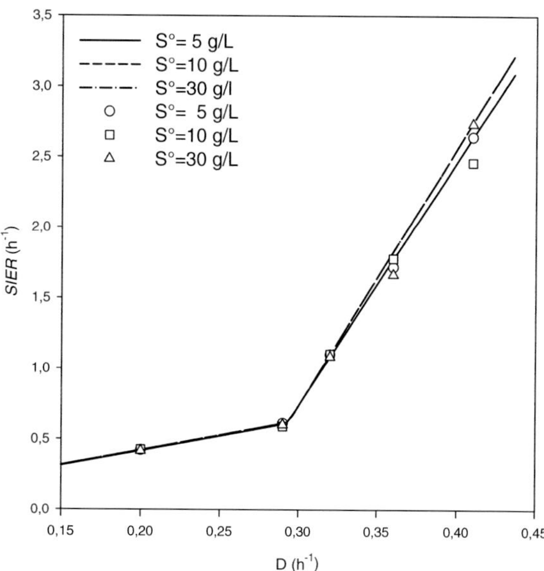

Fig. 3.27 Profile of the specific interphasic exchange rate for the three concentrations at the inlet. *Method* The simulations are done using the algorithm (3.6.24) and the parameters from Table 3.7, with a representation of the substrate done according to the EXP method. *Comments* Practically, the three curves overlap. Note that the dispersion of the experimental results (averages) increases with the dilution rate (After Thierie (2004a,b); permission of Elsevier.)

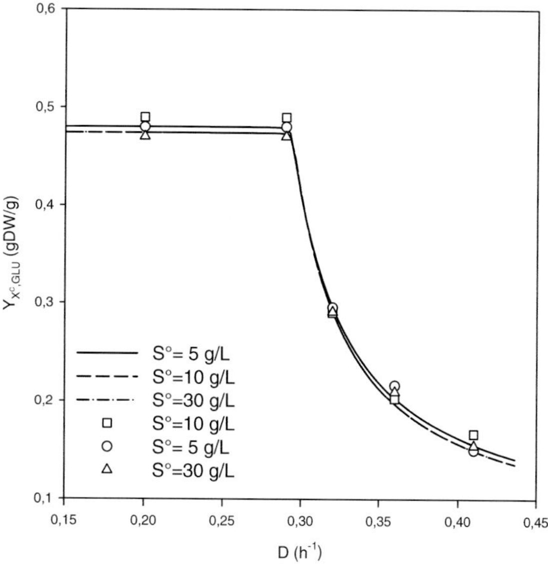

Fig. 3.28 Profile of the yield coefficient for the three concentrations at the *inlet*. *Method* as in Fig. 3.27. *Comments* The three curves overlap and the dispersion of the results has a tendency to increase with D for $D > D_c$ (After Thierie (2004a,b); permission of Elsevier.)

the substrate. This fact, combined with the dependence of K_M on S^0 makes us think that the GM is a less good representation and that the system deviates from the Monod's model, contrary to what certain authors maintained (Sonnleitner and Käpelli 1986).

Remark The EtOH curve for S^0 = 30 g/L is less satisfactory than for the other concentrations at the inlet, especially for $D > 0.35$ h^{-1}. The result can be greatly improved by introducing a correcting term into (3.6.36):

$$q^c_{\mathrm{EtOH}} = \varepsilon\,\rho_{\mathrm{EtOH,GLU}}(1 - \beta)q^c_{\mathrm{GLU}}(l) \tag{3.6.42}$$

The value of ε is of the order of 1.03 and the correction is slight. This way of proceeding (by introducing an arbitrary correction) is not very satisfactory because it disturbs the balance between what is excreted and what remains in the cell phase. Another way of doing this consists of adjusting the curve using the EtOH profile alone. A value for β is then obtained which is equal to about 0.075, (curve in dots in Fig. 3.26). In spite of this new value, the adjustment is not perfect. In addition, the other profiles (biomass, notably) are clearly less good. So the value from Table 3.7 must be considered the best compromise to account of all the curves that were obtained experimentally (and so β is a constant for all concentrations at the inlet). Let us note that the concentrations above are weak, of the order of 15–25 % maximum. Now, β is very sensitive to variations in biomass, for which errors can reach 30 %. So it is not absolutely impossible that this difference can be due to dispersal of the biomass measurements. However, in this case, how can the good agreement of the two other profiles be explained? The experimental point that is most problematic is that situated toward D = 0.41 h^{-1}. This point is very singular because it shows cell production of ethanol that is about twice as high as all other production values. Moreover, production of CO_2 that is associated with the production of ethanol does not present any singularity of this kind (refer to Fig. 3.31). In fact, there are good reasons to think that the problematic experimental point from Rieger et al. could well be an experimental or a typing error (other errors have slipped into the table drawn up, notably concerning certain yield coefficients that are very easy to confirm). Given that each measurement is the average of several points that are quite dispersed, the transcription in the table seems more probable. In conclusion, it is thought that it would be hazardous to construct a correcting hypothesis of one point out of 15 that does not coincide with the model. It would also be unjustified to deduce from a value (perhaps doubtful), a theory of the overproduction of ethanol near the washout point for concentrations of glucose that are raised at the inlet of the chemostat.

Testing of Critical Dilution Rate

It has been shown (refer to (3.6.17)) that the critical dilution rate (threshold) could be approximated by

$$D_C \approx V^0_{\mathrm{GLU}} Y_M$$

Several methods have been used to test this theoretical value. The results appear in Table 3.11.

Method 1 consists of calculating the point of intersection of the two straight lines obtained by SIER (3.6.41) fittings at high and low values of D. Methods 2 and 3 are done in the same way, but using specific production rates of total CO_2 and oxygen, respectively. The agreement is excellent (especially for SIER) and provides a value close to 0.3 h^{-1}.

3.6.2.3 Distribution of Specific Rates in the Cell Phase

General Presentation

The specific rates calculated up to now have been expressed in equivalent substrate, except for ethanol and CO_2 produced by the fermentative pathway. Additional specific rates can be constructed while taking into account that the process is homofermentative. The lacking pathways are those for the production of biomass, respiratory CO_2, and water. A hydrogen consumption pathway should also be added to this. This pathway is necessary to the formation of water. The experimental data is insufficient to treat nitrogen that will not be part of the balance.

The specific rates are conservative that is to say that the sum of the partial specific rates is equal to the total specific rate that is nothing other than the specific interphasic exchange rate (SIER). Figure 3.29 presents the only logical configuration concerning the distribution of partial specific rates in the cell phase.

The glucose penetrates into the cell phase via pathway (E). The nodal point (shaded area in the diagram) serves as a central focus for the fermentative pathway (1), the pathway that preserves a certain quantity of substrate (adsorbed on the membrane or intracellular) (5) and the biosynthesis/fuelling pathway leading to the synthesis of biomass, water, and respiratory CO_2. Inlet pathways for oxygen and hydrogen are necessary for completion of the whole system.

The specific rates to be determined are, therefore, $q^c_{O_2}$, $q^c_{X/c}$, $q^c_{CO_2(r)}$, $q^c_{H_2O}$, and q^c_{H}.

Table 3.11 Calculation of D_c

S^0 (g/L)	D_c			
	Theory	Method 1	Method 2	Method 3
5	0.293	0.294	0.285	0.307
10	0.294	0.294	0.287	0.308
30	0.294	0.294	0.289	0.294

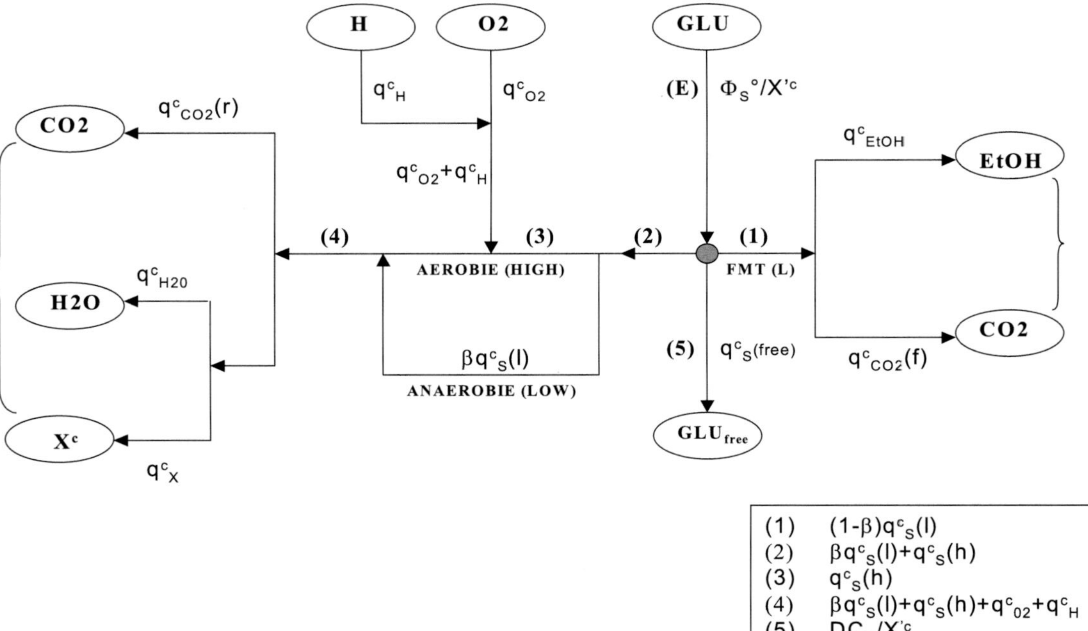

Fig. 3.29 Distribution of the principal specific rates within the cell phase. (*E*) Substrate inlet total specific rate, Φ_S^0/X'^c; (*1*) The fermentative pathway rate (FMT) is a fraction of the low-affinity T/M pathway $(1-\beta)q_S^c(l)$; (*2*) specific rate (in equivalent substrate) of the non-fermentative pathway, including the high-affinity pathway and part of the low-affinity pathway: $q_S^c(h)+\beta q_S^c(l)$; (*3*) High-affinity specific rate (HIGH, in equivalent substrate) associated with oxygen and hydrogen consumption constituting the aerobic pathway that metabolizes glucose (respiratory pathway). (*4*) Sum of metabolism rates that contribute to biosynthesis and associated fuelling, $q_S^c(h)+\beta q_S^c(l)+q_{O_2}^c+q_H^c$; (*5*) Free substrate (adsorbed and/or intracell) rate. The point in the shaded area is called the nodal point and makes up the principal focus for the dispersion of specific rates (expressed in equivalent substrate) (After Thierie (2004a,b); permission of Elsevier.)

Specific Oxygen Consumption Rate

One of the characteristics of the Crabtree effect is to exhibit a typical profile of the specific rate for O_2, close to a function of the Blackman's type (a straight line of positive slope followed by a plateau). The specific rate for the high-affinity pathway shows an identical profile when K_S^* is small enough. It seems, therefore logical to associate the specific oxygen consumption rate with the high-affinity pathway in a simple form such as

$$q_{O_2}^c = \gamma_{O_2} q_{GLU}^c(h) \tag{3.6.43}$$

where γ_{O_2} is a dimensionless constant. The values of this constant have been determined near the critical point and are set out in Table 3.12.

The present constant has a very slight tendency to decrease with concentration at the inlet (the slope is of the order of -10^{-3}). This variation is quite slight so the values can be considered as independent of S^0.

Figure 3.30 shows that the correspondence between (3.6.43) and the experimental data is good. The high-affinity pathway for the metabolism of glucose can reasonably be associated with consumption of oxygen.

Note: A reduction in specific rate seems to be produced for $D > 0.4\ h^{-1}$. An analogous phenomena, but one that is more pronounced, has already been observed (von Meyenburg 1969) and was interpreted as an artifact. (Nielsen and Villadsen 1994). Other studies clearly show a true plateau (Barford and Hall 1979; Barford et al. 1981).

Specific Biomass Production Rate (or Growth)

Due to the conditions for the functioning of the chemostat that are under consideration (those of the biphasic chemostat), the result is that specific production of biomass is

$$q_{X/c}^c = D \tag{3.6.44}$$

This quantity is generally represented by μ in Monod's model.

Table 3.12 Values for γ_{O_2}

S^0 (g/L)	γ_{O_2}
5	0.441
10	0.421
30	0.418

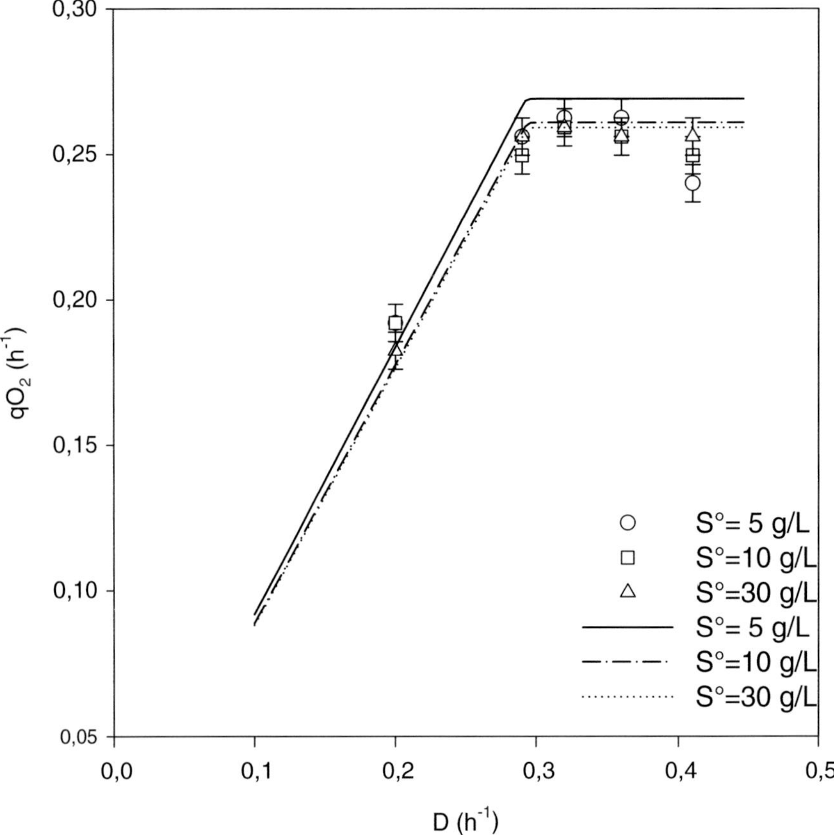

Fig. 3.30 Profile of the O_2 specific consumption rate for the three concentrations at the inlet. *Method* The simulations are done using the quantities derived from the algorithm (3.6.24) and the parameters from Table 3.1, with a representation from the substrate according to the EXP method. $S^0 = 5$ g/L: *open circle* experimental data; *solid line* simulation. $S^0 = 10$ g/L: *open square* experimental data; *dotted and dashed line* simulation. $S^0 = 30$ g/L: *open triangle* experimental data; *dotted line* simulation. The specific rate is calculated according to the relationship (3.6.43). *Comments* A slight dependance according to the inlet substrate seems to appear at the plateau ($D > D_c$). It is, however, very low and the curves could be considered as confused (refer to text and Table 3.12) (After Thierie (2004a,b); permission of Elsevier.)

Specific Water Production Rate

In the respiratory chain of reactions, there is an empirical ratio between the O_2 consumption rate and the water (H_2O) production:

$$q^c_{H_2O} = \gamma_{H_2O} q^c_{O_2} \tag{3.6.45}$$

Table 3.13 gives the values from this estimation.

Table 3.13 γ_{H_2O} values

S^0 (g/L)	γ_{H_2O}
5	0.89
10	0.90
30	0.90

Specific Hydrogen Consumption Rate

Consider equation,

$$O_2 + 4H \rightarrow 2H_2O \tag{3.6.46}$$

and use the relationship

$$q_H^c = \rho_{H,O_2} q_{O_2}^c \tag{3.6.47}$$

The mass coefficient, equal to 0.125 makes it possible to calculate the specific rate for hydrogen consumption associated with the consumption of oxygen for water production.

Note If all the oxygen was consumed to produce water, the empirical constant γ_{H_2O} would be equal to the mass coefficient ρ_{H_2O,O_2} that is 1.125. Instead of this value, there is 0.90 (refer to Table 3.11), which signifies that about 80 % of the oxygen is used to produce water. The remaining 20 % is therefore used to form oxidized compounds of another nature. (Obviously, these include oxidation of nitrogen, unmeasured in experiments of Rieger et al.) Equation (3.6.46) is therefore incomplete and can only be used for hydrogen.

Specific Respiratory CO_2 Production Rate

Conservation of specific rates allows to write (cf. Fig. 3.29)

$$q_{CO_2(r)}^c = q_{GLU}^c(h) + \beta q_{GLU}^c(l) + q_{O_2}^c + q_H^c - (q_{H_2O}^c + q_{X/c}^c) \tag{3.6.48}$$

That makes it possible to deduce the specific production rate of respiratory CO_2 from other rates. Figure 3.31 shows the good agreement between specific production rate of total CO_2 and the values calculated via (3.6.39) and (3.6.48) (CO_2(f) + CO_2(r)). This agreement justifies the method of calculation of $q_{CO_2(r)}^c$ and also all the other necessary rates used to calculate this.

Figure 3.32 shows the values for specific rates just before ($D = 0.28\ h^{-1}$; higher values, in bold) and just after ($D = 0.32\ h^{-1}$; lower values in italics) the critical dilution rate. The change between the respiratory regime and the respiro-fermentative regime is obvious. It is easily verified that the specific rates (S.R.) are preserved and that the balance of the S.R. at the inlet compensates for the

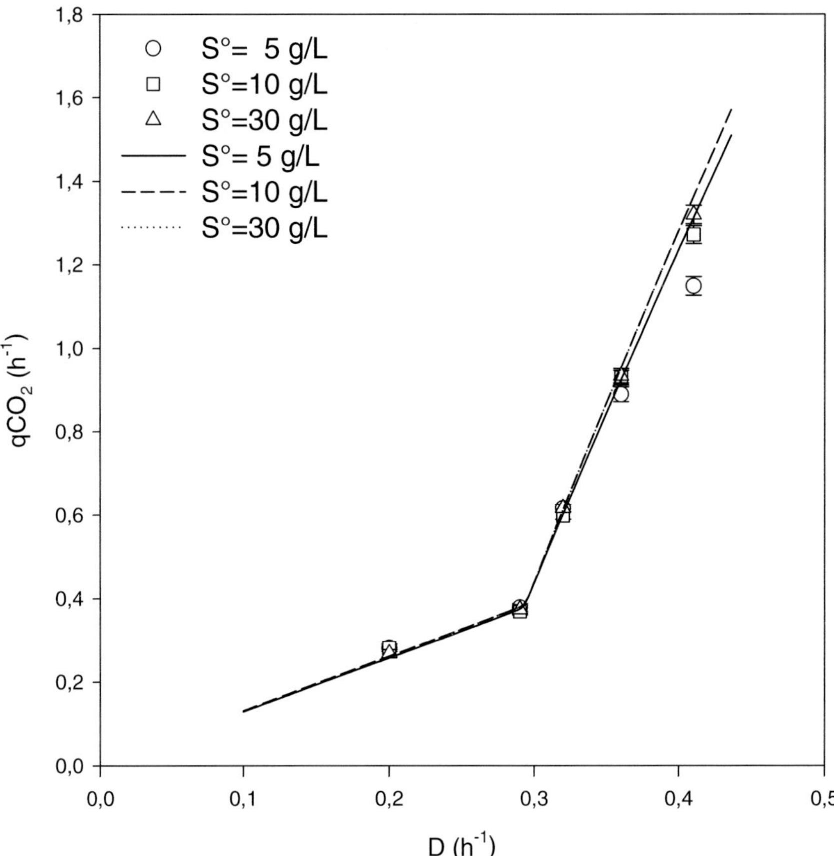

Fig. 3.31 Profile of the specific production rate of total CO_2 for the three concentrations at the inlet. *Method* as in Fig. 3.30. The total specific rate (respiration + fermentation) is calculated by the sum of the relationships (3.6.39) and (3.6.48). *Comments* The three curves overlap and the dispersion of the average experimental data increases with D. Note that there is no increase in the production rate of CO_2 for $D > 0.32\ h^{-1}$ for $S^0 = 30$ g/l. This seems contradictory to significant increase in concentration of ethanol for these values for dilution rate (Refer to Fig. 3.26) (After Thierie (2004a,b); permission of Elsevier.)

balance of the S.R. at the outlet. (This verification can be done both globally (the entire diagram) and following specific pathways (large curly brackets).

3.6.2.4 Checking Mass Balance at the Molecular and Elemental Levels

Rieger et al. (1983) have determined the elemental composition of the biomass and have obtained the following balance:

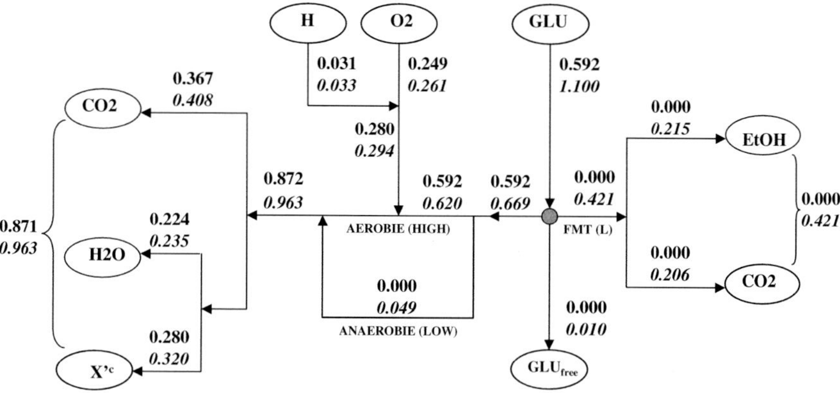

Fig. 3.32 Quantification of the principal specific rates within the cell phase. *Method* The calculations have been done with kinetic parameters from Table 3.7 and the substrate has been evaluated using the EXP method. The concentration at the inlet is S^0 = 10 g/L. *Comments* This diagram shows numerical values corresponding to equation from Fig. 3.29. The values have been calculated just before and just after the critical value D_c = 0.3 h^{-1}. The higher value (*in bold*) corresponds to $D = 0.28$ h^{-1}; the lower value (*in italics*) to $D = 0.32$ h^{-1}. It is noted that the specific rates ANAEROBIC (LOW), of fermentation (FMT(L)), as well as the pathway for free substrate (adsorbed and/or intracellular) are zero before the critical value of 0.3 h^{-1}. Conservation of the specific rates can be globally (sum of the inlet flows = sum of the outlet flows) or partially (curly brackets) checked. To calculate the metabolic flows, just multiply the values of the specific rates by the biomass ($X^{/c}$) corresponding to the dilution rates used (After Thierie (2004a,b); permission of Elsevier.)

$$C_6H_{12}O_6 + \alpha_1 O_2 + \alpha_2\, 0.15NH_3 \rightarrow \alpha_2 CH_{1.79}O_{0.57}N_{0.15} + \alpha_3 C_2H_6O + \alpha_4 CO_2 + \alpha_5 H_2O \tag{3.6.49}$$

They were in a position to calculate the values of the different stoichiometric coefficients for the experimental values of D. (Note that nitrogen, for which there is no experimental data, is eliminated from the equation.) These results can be used to check, by a completely independent method, whether the different specific rates that we have calculated and the model, in general both verify the elemental balance (3.6.49). It suffices to check whether the values of the stoichiometric coefficients are comparable to those of the authors. In the text that follows, the molecular mass of the biomass is used when it equals that appearing in (3.6.49), that is to say 25 atomic mass units (amu).

Using (3.6.26) and (3.6.28), it can be demonstrated that

$$v_{Pi} = \frac{q^c_{Pi}}{\mathrm{MM}_i} \quad \frac{v_S \mathrm{MM}_S}{q^c_S} \tag{3.6.50}$$

where ν_j and MM_j are the stoichiometric coefficients and the molar mass of compound j.

In the special case (3.6.49), the following is found:

$$\nu_{Pi} = \frac{q^c_{Pi}}{\mathrm{MM}_i} \frac{180}{q^c_{\mathrm{GLU}}(E)} \tag{3.6.51}$$

with $q^c_{\mathrm{GLU}}(E) = \Phi^0_{\mathrm{GLU}}/X^{/c}$, the specific interphasic exchange rate (SIER). (In fact, it is a question of all the glucose implied in the reaction (3.6.49), which excludes the residual glucose in the chemostat.)

So this relationship makes it possible to calculate all the stoichiometric coefficients and also the molecular balance and elemental mass, on condition that $\mathrm{MM}_{X/c}$ is known. Table 3.14 shows the comparison between the coefficients obtained experimentally by Rieger et al. (α), and those calculated by (3.6.51), (ν). The c_j values at the head of the columns denote the indices of the coefficients such as they appear in (3.6.49). The column on the right (MB) shows the difference (in percentage) with the exact mass balance. The approach makes it possible to differentiate the stoichiometric coefficients of CO_2 using the respiratory pathway ($\nu_4(\mathrm{resp})$) or the respiro-fermentative one ($\nu_4(\mathrm{fer})$), with, evidently $\nu_4 = \nu_4(\mathrm{resp}) + \nu_4(\mathrm{fer})$.

On the whole, the agreement of the results is good and the mass balance is always higher than 96 % which is perfectly compatible with the experimental data. However, it should be noted that the mass balance tends to rise with D, which probably indicates a slight systematic bias. This bias could be explained by simplification of the representation of the process (3.6.49) and notably through the absence of data concerning the nitrogen fluxes that represent 8 % of the biomass. Another cause could stem from the absence of free glucose in (3.6.49), in which case a corrective term should be brought into (3.6.51). The differences from the ideal are too slight and the experimental data are insufficient for correction to be relevant and it can be agreed that the results obtained are satisfactory.

3.6.2.5 Approximate Form of the Excretion of Metabolites

Given here is a simplified approximate form that makes it possible to calculate the metabolite concentration in the matrix phase without having to calculate all the specific rates. So part of the parameter estimate of the process is avoided. This can easily be done by hand if the experimental points are available $\{S, X^{/c}\}$ as well as the yield coefficient, Y_M (obtained for $D < D_c$). However, the method is only available if there is no more than a little free glucose in the cell phase. The interest in the method lies in the fact that it is possible to estimate (a) the theoretical production of a metabolite produced by homofermentation, without having to measure it directly; (b) test whether the process is really homofermentative.

Table 3.14 Comparison of the stoichiometric coefficients. $C_6H_{12}O_6 + c_1.O_2 + c_2.0.15.NH_3 \rightarrow c_2.CH_{1.79}O_{0.57}N_{0.15} + c_3.C_2H_60 + c_4.CO_2 + c_5.H_2O$

D (h^{-1})	S^0 (g/L)		c_1	c_2	c_3	c_4	c4 resp	c4 fer	c_5	MB (%)
0.2	5	α	2.53	3.30	0	2.70	–	–	3.79	
		ν	2.48	3.46	0	2.55	2.55	0	3.92	−0.4
	10	α	2.52	3.31	0	2.69	–	–	3.78	
		ν	2.37	3.41	0	2.54	2.54	0	3.79	−0.3
	30	α	2.45	3.38	0	2.62	–	–	3.74	
		ν	2.35	3.41	0	2.54	2.54	0	3.76	−0.2
0.29	5	α	2.41	3.42	0	2.58	–	–	3.71	
		ν	2.47	3.44	0	2.55	2.55	0	3.91	−0.4
	10	α	2.37	3.46	0	2.54	–	–	3.68	
		ν	2.36	3.40	0	2.54	2.54	0	3.77	−0.3
	30	α	2.41	3.42	0	2.58	–	–	3.71	
		ν	2.34	3.40	0	2.53	2.53	0	3.75	−0.3
0.32	5	α	1.33	2.04	0.84	2.28	–	–	2.11	
		ν	1.40	2.13	0.76	2.29	1.52	0.77	2.21	0.6
	10	α	1.34	2.08	0.83	2.20	–	–	2.12	
		ν	1.33	2.09	0.77	2.28	1.51	0.77	2.13	0.7
	30	α	1.33	2.05	0.84	2.27	–	–	2.11	
		ν	1.32	2.09	0.77	2.28	1.51	0.77	2.12	0.7
0.36	5	α	0.87	1.50	1.18	2.13	–	–	1.45	
		ν	0.85	1.46	1.13	2.14	1.01	1.13	1.35	2.0
	10	α	0.80	1.41	1.24	2.11	–	–	1.34	
		ν	0.80	1.42	1.14	2.13	0.99	1.14	1.28	2.1
	30	α	0.80	1.41	1.24	2.11	–	–	1.34	
		ν	0.80	1.42	1.14	2.13	0.99	1.14	1.27	2.1
0.41	5	α	0.59	1.26	1.35	2.50	–	–	1.01	
		ν	0.57	1.12	1.30	2.04	0.74	1.30	0.91	3.7
	10	α	0.55	1.13	1.42	2.03	–	–	0.98	
		ν	0.53	1.08	1.32	2.03	0.72	1.32	0.86	3.8
	30	α	0.54	1.09	1.44	2.03	–	–	0.95	
		ν	0.53	1.08	1.32	2.03	0.72	1.32	0.85	3.8

It has previously been shown that

$$\frac{X^{/c}}{X^c} = \frac{q_S^c(h) + \beta q_S^c(l)}{q_S^c(h) + q_S^c(l)} \leq 1 \tag{3.6.52}$$

This relationship makes it possible to calculate the difference $X^c - X^{/c} \geq 0$:

$$X^c - X^{/c} = \frac{X^c}{q_S^c(h) + q_S^c(l)}(1-\beta)q_S^c(l) \tag{3.6.53}$$

When $\tilde{C}_S^c \ll 1$, preservation of the specific rates (in equivalent substrate) can be approximated by the following relationship (cf. Fig. 3.29):

$$q_S^c(h) + q_S^c(l) \approx \frac{\Phi_S^0}{X^{/c}} \tag{3.6.54}$$

By inserting this value into (3.6.53), this equation is formed,

$$X^c - X^{/c} = \frac{X^c}{\Phi_S^0} X^{/c}(1-\beta)q_S^c(l) \tag{3.6.55}$$

By multiplying the left-hand side and the right by the mass coefficient of the metabolite P_i and by comparing the right side with (3.6.31), it is obvious that

$$\tilde{C}_{P_i}^m = (X^c - X^{/c})\frac{\rho_{P_i,S}}{Y_M} \tag{3.6.56}$$

X^c can be calculated from the experimental data as in the algorithm (3.6.24), $X^c = Y_M(S^0 - S)$. For all experimental values $X_{\text{exp}} = X^{/c}$ it is possible to calculate the concentration of the metabolite P_i (provided that the biomass is not too great).

The relationship (3.6.56) has been compared with the thorough method used above by putting

$$\theta_{P_i} = \frac{\rho_{P_i,\text{GLU}}}{Y_M}\ ;\ P_i = \text{EtOH, } CO_2 \tag{3.6.57}$$

The fittings in the experimental method (EXP) and by the generating model method (GM) give the same results for all concentrations at the inlet. Comparison of the thorough methods and the approximate Eq. (3.6.57) is given in Table 3.15. (The value of θ in (3.6.57) has been obtained by manual adjustment, associated with calculation of least squares, in such a way that $P_i(D)$ is as close as possible to the thorough fittings.)

Table 3.15 Comparison of the thorough methods and the approximation (3.6.56)

S^0 (g/L)	θ_{EtOH} (EXP, MG)	θ_{EtOH} (3.6.57)	θ_{CO_2} (EXP, MG)	θ_{CO_2} (3.6.57)
5	1.05	1.06	1.00	1.02
10	1.05	1.08	1.00	1.03
30	1.05	1.08	1.00	1.03

It is clear that in order to obtain a simple estimate, Eq. (3.6.56) gives results that are completely compatible with those given by the complete model. It is doubtful whether a more thorough parameter estimate procedure would radically change the result.

3.6.3 Discussion

Several types of models or representation of metabolic flow use optimization criteria (refer to Sect. 3.5) as a constraint to resolve an underdetermined algorithm. This is the case, for example, of systems having more unknowns than equations. These criteria are often based on very general principles that are often not demonstrated. In fact, it is often assumptions based on concepts linked to the theory of biological evolution. Among these criteria, maximization of growth rate reappears repeatedly (Varma and Palsson 1993a, b, 1994; Bellgardt 2000b). In a chemostat, this criteria cannot be used because the growth rate is dictated by conditions external to the culture and the growth rate is not maximum but equal to the dilution rate ($\mu = D$). Obviously, this relationship can be used as a constraint, but it is at a global level, of description that is badly adapted to calculation of metabolic flows. This approach makes it possible to show an optimization criteria that applies when certain conditions are fulfilled.

Let us define the specific metabolism rate using (refer to Fig. 3.29)

$$q_S^c(\text{met}) = q_S^c(h) + \beta q_S^c(l) \tag{3.6.58}$$

This rate characterizes the whole anabolism processes (biosynthesis) and the associated fuelling reactions.

Moreover, the specific interphasic exchange rate (SIER) corresponds to the net total rate of equivalent substrate that crosses the cell phase,

$$\text{SIER} = q_S^c(h) + q_S^c(l) + q_S^c(\text{free}) = \frac{\Phi_S^0}{X/c} \tag{3.6.59}$$

where $q_S^c(\text{free})$ is the specific rate of the free (nonmetabolized) substrate. The ratio of the specific metabolization rate and the total specific rate is then given by

$$\frac{q_S^c(\text{met})}{\text{SIER}} = \frac{q_S^c(h) + \beta q_S^c(l)}{q_S^c(h) + q_S^c(l) + q_S^c(\text{free})} \tag{3.6.60}$$

Before the switch ($D < D_c$), $q_S^c(l) = q_S^c(\text{free}) = 0$ and so

$$q_S^c(\text{met}) = \text{SIER} = \frac{D}{Y_M} \tag{3.6.61}$$

The metabolic rate is then at its maximum.

After the switch, if $q_S^c(\text{free})$ is negligible in comparison with other specific rates, (3.6.60) becomes

$$\frac{q_S^c(\text{met})}{\text{SIER}} = \frac{q_S^c(h) + \beta q_S^c(l)}{q_S^c(h) + q_S^c(l)} \tag{3.6.62}$$

It has been demonstrated earlier that (refer to Sect. 3.5),

$$X^{/c} = \frac{q_S^c(h) + \beta q_S^c(l)}{q_S^c(h) + q_S^c(l)} X^c \tag{3.6.63}$$

Using this relationship and (3.6.59), the relationship (3.6.62) becomes

$$\frac{q_S^c(\text{met})}{\Phi_S^0} = \frac{1}{X^c} \tag{3.6.64}$$

By rearranging and using the definition of the yield coefficient, it is finally obtained that

$$q_S^c(\text{met}) = \frac{D}{Y_M} \tag{3.6.65}$$

This is the same value as (3.6.61).

In other terms, the metabolic rate keeps its maximum value both in the respiratory regime and in the respiro-fermentative regime on condition that the low-affinity pathway can (k_0^* quite big) treat enough substrate to allow the intracellular substrate concentration to remain low. This is effectively what happens in the example from Rieger et al. we have studied. For all analyses of the underdetermined metabolic fluxes concerning systems of this type, the relevant constraint would not be maximization of the growth rate, but the maximization of the net metabolic specific rate (or fluxes).

After the switch, there is a "surplus" specific rate

$$q_S^c(\text{exc}) = \frac{\Phi_S^0}{X^{/c}} - \frac{D}{Y_M} \tag{3.6.66}$$

that corresponds to the specific rate of excretion of metabolites (ethanol and fermentative CO_2). In order to have metabolites excretion, it is evident that the capacity for substrate transport must be able to exceed the maximum metabolic rate. This agrees with observations from Van Urk et al. (1989). The model does not make it possible to know if the transport of the surplus substrate is regulated so that the excreted flow is adapted to the maximum metabolism (to ensure a redox balance, for example) or if on the contrary the metabolic rate acts as a bottleneck that causes excretion of a surplus of substrate that the cell cannot avoid. This last interpretation is a return considering that no regulated inflow of substrate would make it possible for glucose to serve as a signal for (a) saturating the high-affinity pathway and (b) triggering the low-affinity pathway. This signal would then have an effect that was purely kinetic, since then the concentration increases (after the threshold) the hyperbolic form and saturates when the reaction of order 1 becomes significant. The Crabtree effect could then be interpreted in terms of competition between the two metabolic pathways, regulated by the substrate. There are arguments in favor of regulation at the inflow of substrate (Van Urk et al. 1989; Lagunas et al. 1982; Lagunas 1993; Carlson 1987). To the authors' knowledge, it is not possible to confirm at present that regulation is always effective for all transport systems that can represent a strain (Walsh et al. 1996). Recent studies (Cortassa and Aon 1997, 1998) seem to confirm the regulation by substrate approach that is in play at the level of transport pathways and metabolic pathways at the same time.

An interesting point can also be drawn from (3.6.61) and (3.6.65). By combining (3.6.44) with one of these relationships, it is easily shown that

$$\frac{q^c_{X/c}}{q^c_S(\text{met})} = Y_M \tag{3.6.67}$$

This relationship shows that the ratio of specific biomass production rate (or flux) and the metabolic (biosynthesis + fuelling) specific rate (or flux), expressed in equivalent substrate, is constant before and after the switch. Moreover, this value is nothing other than Y_M, the "yield coefficient" of the generating model. As it is not a question of a true yield coefficient of the system, it is suggested to call this the biosynthetic efficiency coefficient. This coefficient can be considered as a measurement of the coupling of the biosynthesis flow and the associated fuelling flow. This result deserves to be confirmed in other cases, but the value obtained in this special case shows that the coupling of two pathways (expressed in equivalent substrate) remains invariant and has its maximum value for the respiratory regime as well as the respiro-fermentative regime.

In conclusion, the example of the Crabtree effect in *S. cerevisiae* shows that a model with two transport/metabolism pathways makes it possible to take into account the main properties of a sudden change from a respiratory regime to a respiro-fermentative regime in the case of homofermentative production of ethanol. A kinetic that is near to (3.6.9) in its local form has been used efficiently by Walsh et al. (1994) to describe the glucose transport rate in the wild-type strain X-2180 of *S. cerevisiae* cultivated in batch. The low-affinity component (of first order or nonsaturable) is interpreted by the authors as a simple diffusion term. Moreover, using the model with two pathways, they note that the affinity for glucose decreases when the biomass increases. The data from the article are not sufficient to verify whether the product of the half-saturation coefficient by the biomass is constant, but the variation in affinity observed by the authors goes very much in the same direction as that given by the global form ($K^*_S X^{/c}$) that appears in (3.6.9). These results significantly confirm the idea that the global representation is justified.

From the point of view of metabolism, it has been shown that the kinetic parameters can be considered as true constants over the entire interval of dilution rates and for three concentrations at the chemostat inlet. The main advantage of the model is that in addition to its great simplicity, it does not require optimization criteria or logical tests (Sonnleitner and Käpelli 1986) defined a priori for the good functioning of the algorithm. The hyperbolic form of the kinetics well reflects the transport by diffusion or facilitated diffusion phenomenon that characterizes the transport of glucose in *S. cerevisiae* (Walker 1998; Schechter 1997). The stoichiometric mass balance confirms the validity of the mass balances for the PDS using an independent method.

This representation rests explicitly on the fact that the whole process is governed by interphasic exchange substrate phenomena and on intracellular metabolism of this. In particular, the specific rate of consumption of oxygen is deduced from the high-affinity metabolic pathway. This makes it possible to take into account correctly, the specific rate of O_2 and also to check the stoichiometry of the reaction. These results favor an explanation of the Crabtree effect based on metabolism and transport of glucose and not on the limitation of the capacity of the respiratory chain reactions (Sonnleitner and Käpelli 1986). Different arguments that move in the same direction can be found in Walker (1998).

From the point of view of biotechnologies, it is thought that the kinetic parameters of the model make it possible to characterize quantitatively a strain that is engaged in a determined process. So for example, it is possible to quantify the effects of a genetic modification of a strain or evaluate the impact of the modification of a culture medium. In particular, the relationship that links the critical dilution rate to intrinsic, global features of a culture ($D_c = V_S^0 Y_M$) could serve as a quantity guide for optimization of the process aimed at producing preferentially biomass (such as baker's yeast) or rather metabolites (such as ethanol) in the fermentation industry. Cortassa and Aon (1998) have, for example, shown that the value of D_c can vary according to mutants that cause alterations genes (disruptions) that are repressed by glucose (glucose-repressible). In this study, it appears that all mutants have a critical dilution rate and a maximum specific T/M rate that are slower than those of the wild strain. Except for mutant (*snf4*), all the yield coefficients are also lower. This last example shows that it is possible to preserve a global quantity such as yield by modifying a gene at the same time as modifying the maximum high-affinity specific T/M rate (in the case of *snf4*, the reduction is of the order of 25 %). Inversely, it is also possible to modify the global quantities and local quantities simultaneously. This approach makes it possible to quantify easily this type of modification and so provide an evaluation and optimization tool.

3.7 Respiro-Fermentative Phenomena in Bacterial Flocs

3.7.1 Introduction

In natural (or seminatural) media contrary to the case in the last section, free (isolated) bacteria only represent a very small part of the population. The great majority of these microorganisms (more than 90 %, or even 99 %, Costerton et al. 1995) live in the form of aggregates, biofilms, and flocs. It is generally conceded that these structures are ecologically advantageous to individuals. The ecological and public health importance has led to a growing number of studies on the subject.

Two types of major themes are approached. On one hand interest is taken in modifications of specific properties, as the properties of an isolated strain are not necessarily the same when this species is part of a biofilm [for example, resistance to antibiotics and to decontaminating agents (Costerton and Lappin-Scott 1989)].

On the other hand, recent discoveries concerning communication of microorganisms among themselves (quorum sensing, signaling; Shapiro 1991) have aroused great enthusiasm among microbiologists.

Moreover, the concept of consortium (that is not necessarily linked to biofilm) underpins the idea of interspecific collaboration in the accomplishment of a common objective (such as xenobiotics degradation, for example).

In this section, attempts are made to link the two concepts (flocs and consortium) in a perspective that is both fundamental and applied, using PDS theory and previous results.

On one hand, attempts are made to understand and describe certain metabolic characteristics that play a part in the core of a floc (in this case, two types of fermentation or respiro-fermentation). On the other hand, the potential advantage of the use of flocs from the perspective of industrial production is shown. Continuous culture (such as the chemostat) can in effect present economic advantages (Nielsen and Villadsen 1994) compared with other types of culture (batch or fed-batch). One of the snags in the development of these processes of continuous culture is their relative fragility in relation to contamination by external microorganisms. It is shown here that the commercial production of organic acids can be done by a bacterial consortium "packaged" in flocs that are perfectly stable and naturally resistant to contamination. Continuous production is therefore possible away from any sterilization of the bioreactor provided that the work is done on flocs that were carefully selected and acclimatized.

3.7.2 *Materials and Methods*

The results were obtained in our laboratory and are described in detail in (Bensaïd 2000). Here the principal data are taken up again to facilitate understanding.

3.7.2.1 Initial Inoculum

The initial inoculum is provided by the wastewater treatment plant at Erasme Hospital (ULB CHU, Brussels, Belgium). The activated sludge was acclimatized in the chemostat for about 2 years before forming the stable consortium that was used for experiments.

3.7.2.2 Chemostat

A chemostat setup in our laboratory was used. The useful volume was 1 liter with a gravity outlet for effluent. A precision peristaltic pump (Watson-Marlow 603S) was used to regulate the inlet volumic flux. The culture was accomplished in an aerobic medium by maintaining oxygen saturation higher than 50 %. Neither the pH nor any other parameter, were regulated. No precaution whatsoever as regards sterilization was taken within the vessel of the reactor (only the feeding substrate was sterilized). After interruption of the culture, all experiments were repeated with an inoculum provided by the last culture, possibly conserved by freezing. The walls of the reactor were carefully cleaned in order to completely eliminate the biofilm and to ascertain that the obtained results were due only to suspended biomass.

3.7.2.3 Culture Medium

The culture medium was achieved by the improvement of the synthetic sewage feed (SSF) that was allowed by guideline 209 of the OECD guidelines (OECD 1991). This complex substrate (meat extract and peptone) was modified by addition of glucose to obtain a final concentration of 0.5 gGLU/L.

3.7.2.4 Analytical Methods

Biomass was determined by optical density at 600 nm or by weighing at constant weight after drying at 103 °C. The dissolved oxygen was measured by an oxymeter WTW E096. The glucose and the acids were measured by HPLC according to the methods described by Senn et al. (1994) and Van Urk et al. (1988).

3.7.2.5 Bacterial Identification

Enterobacteriaceae and Gram-negative bacteria were characterized and identified by method API 20 E (BioMérieux); the anaerobic bacteria, by the API 20 A method. The counting was done on Petri dish using the method described by Bensaïd (2000).

3.7.3 Results

3.7.3.1 Steady States

The data presented are the values of the culture in steady state (Refer to Bensaïd (2000) for testing the steady state). This means that for each value for dilution rate, the various measured parameters remain constant over time. In particular, the specific rate of production of biomass in the form of flocs compensates exactly for the hydraulic outlet flow of the chemostat. If μ is the specific production rate for the flocs, the relationship $\mu = D$ is verified. The dilution rate, therefore, appears as an external control parameter that makes it possible to fix the production rate for biomass by modulating the volume flow.

3.7.3.2 Biomass, Residual Glucose, and Fermentation Products

The principal results appear in Figs. 3.33 and 3.34.

Figure 3.33 shows that at lower dilution rates, D, ($D = Q/V$ where Q is the volume flow at the inlet and V_T the useful volume of the chemostat) the biomass remains about constant, the residual glucose is undetectable and the lactic acid is absent from the medium.

For a dilution rate that is greater than about 0.25 h^{-1}, biomass decrease rapidly with D, the residual glucose and lactic acid appear in the medium. Note that the

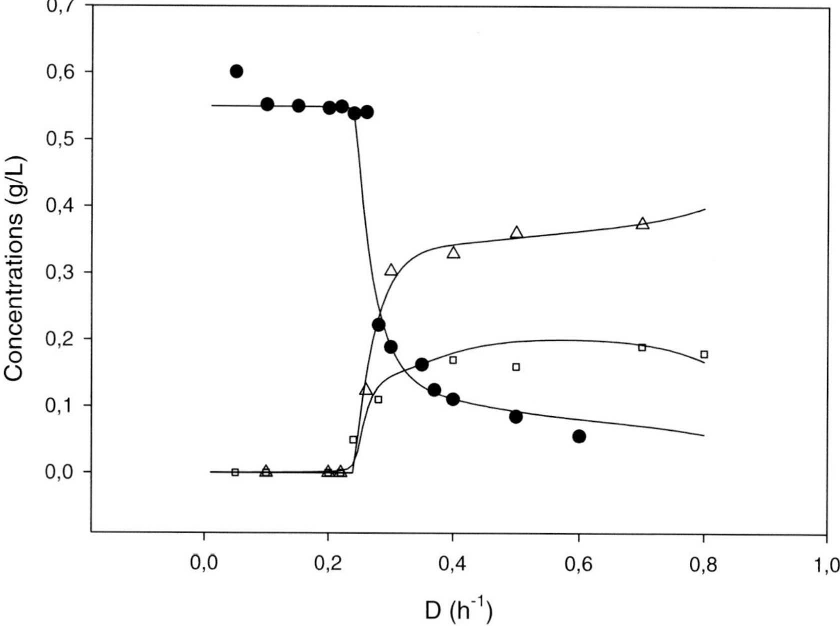

Fig. 3.33 Biomass, residual glucose, and lactic acid according to dilution rate. The figure shows the experimental concentrations at steady state of biomass (*black circle*), residual glucose (*black triangle*), and lactic acid (*black square*) according to dilution rate. The *solid lines* are the result of the adjustment of residual glucose or of simulation of biomass and lactic acid by the model. The parameter estimate has provided the following values: $K_S^* = 1.10^{-6}$; $V_S^0 = 0.58\ h^{-1}$; $K_0^* = 250$; $\beta = 0.36$; $k = 1.5$; the following values are data: $\tilde{C}_S^{m,E} = 1.3$ g/L; $Y_{X^c,S} = 0.423$

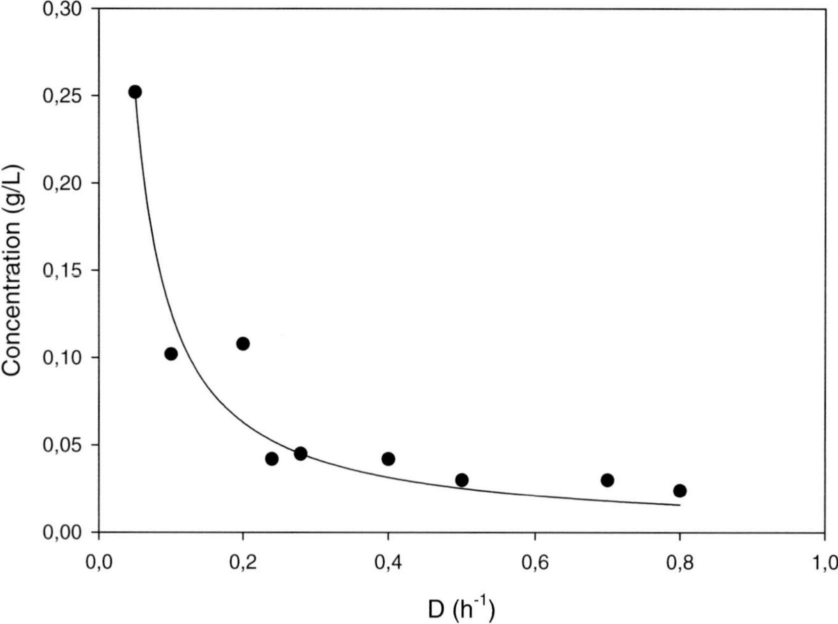

Fig. 3.34 Butyric acid according to dilution rate. The *solid circles* show the experimental values and the solid line the adjustment to the model represented by Eq. (3.7.2) (r^2 = 0.87). The parametric estimate gives a value of $\Pi^m_{\mathrm{BUT}} = 0.012 \pm 0.001$ gBUT/L

profile of the glucose versus dilution rate does not correspond to the profile observed in Monod's model (refer to Sect. 3.2). Figure 3.34 shows the profile for butyric acid according to *D*.

The results are quite dispersed, but show a net tendency to decrease. Contrary to lactic acid, butyric acid is present all through the interval for *D*. These results are perfectly reproducible.

It is the sudden discontinuity in the observed biomass profile in Fig. 3.33 that attracted attention and so caused efforts to be made to understand the underlying mechanism.

3.7.3.3 Pulse of Substrate and Concentration of Dissolved Oxygen

At dilution rates below the critical value, a rapid addition of substrate to the medium (a pulse) leads to the rapid reduction in dissolved oxygen in the medium. Beyond the critical value, this phenomenon does not occur (concentration of dissolved oxygen remains constant). (See Figs. 3.35a, b.)

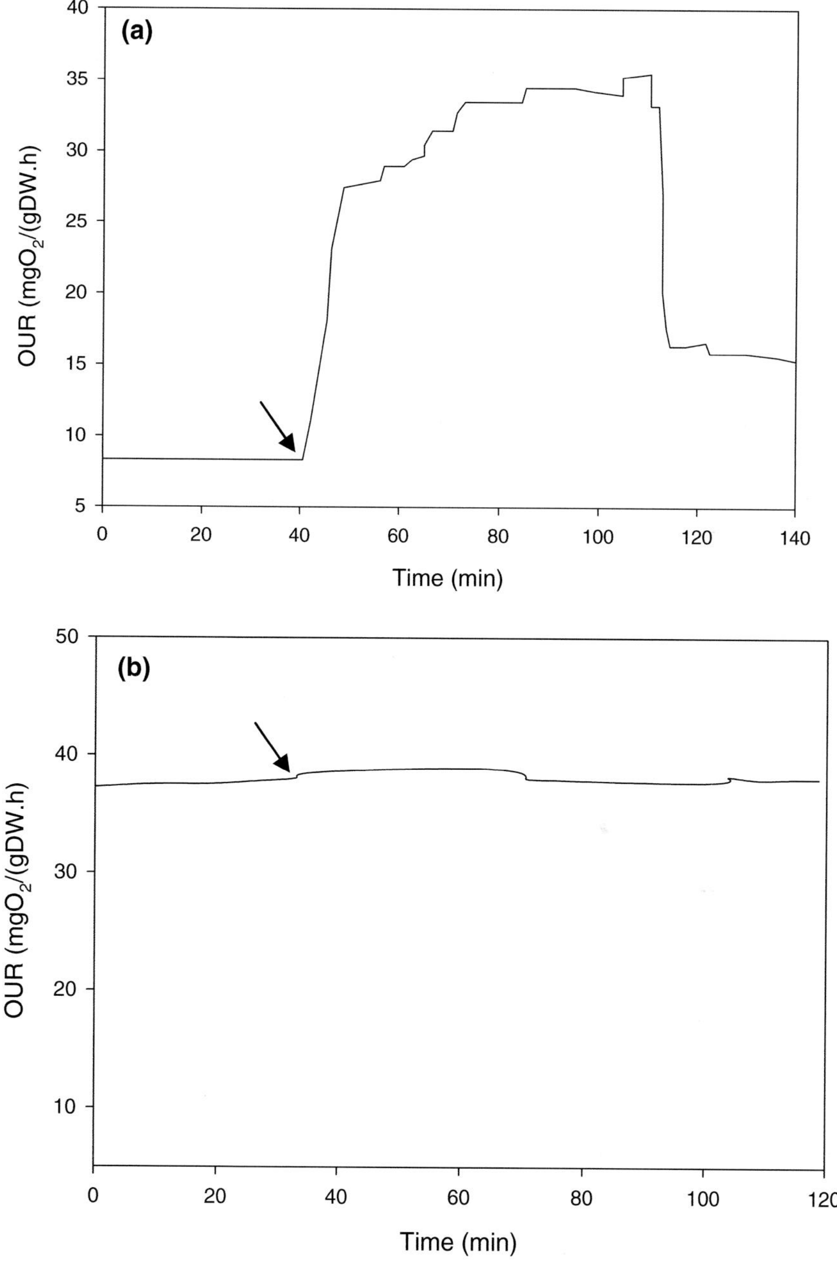

Fig. 3.35 Oxygen uptake rate (OUR) in function of time. In the steady state, 50 mL of substrate (SSF + glucose) were added to the medium (*arrow*). For $D < D_c$ (case a), a sudden increase in the OUR occurs. The specific oxygen uptake rate finally goes back to the steady state preceding the pulse. For $D > D_c$ (case b), the OUR remains roughly unchanged

3.7.3.4 Composition of the Consortium

Using PCR, it was demonstrated that no yeasts were present in the consortium. Seven species of bacteria were identified—*E. coli*, *Lactobacillus plantarum*, *Clostridium novyi*, *Enterobacter agglomerans*, *Enterobacter aerogenes*, *Pseudomonas aeruginosa,* and *Proteus vulgaris*.

They constitute more than 95 % of the total population. The first two form, equally, from 60 to 70 % of the population, the third about 15 %, the four others share the rest in an approximately equivalent manner (in the order of 5 %). The species that have been determined, as well as their relative proportions, remain identical before and after the critical dilution rate. Following these results, the sudden biomass decrease cannot be interpreted in terms of reorganization of the consortium (by the disappearance of a dominant species, for example) and another mechanism should intervene.

Figure 3.36 shows the composition of the consortium before and after the transition.

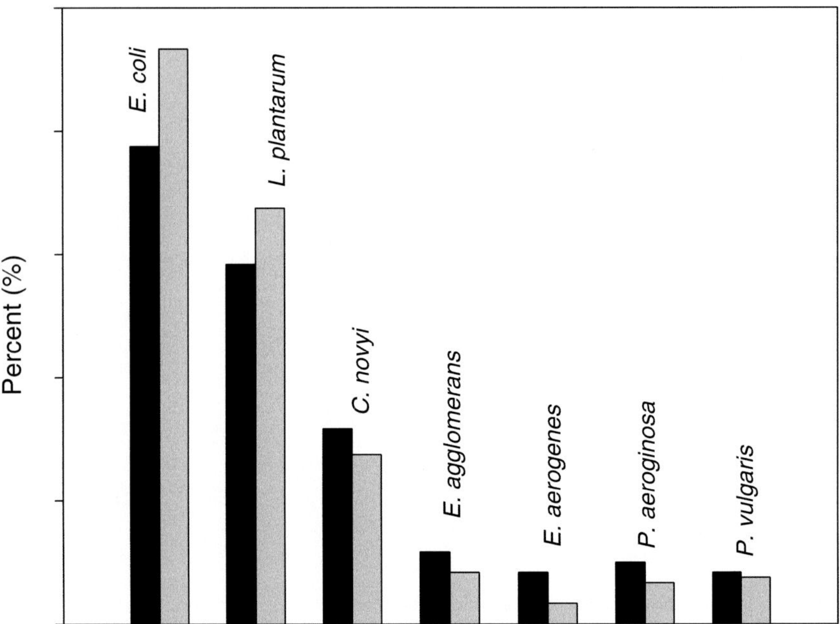

Fig. 3.36 Relative composition of the consortium. The figure shows the relative composition of the consortium at a dilution rate below (*black bars* $D = 0.1\ h^{-1}$) and above (*gray bars* $D = 0.4\ h^{-1}$) the critical value

3.7.4 Modeling

3.7.4.1 Butyrate

The mass balance in the matrix phase (the culture medium) is

$$\frac{d\tilde{C}^{m}_{\mathrm{BUT}}}{dt} = \Pi^{m}_{\mathrm{BUT}} - D\tilde{C}^{m}_{\mathrm{BUT}} \tag{3.7.1}$$

where $\tilde{C}^{m}_{\mathrm{BUT}}$ is the pseudo-homogeneous concentration (for more details of this concept, refer to Chap. 2) of butyrate in the medium. Given the low concentration of biomass, no correction whatsoever is necessary and the pseudo-homogeneous concentration can be considered as equal to the experimental (measured) concentration. Π^{m}_{BUT} is the net butyrate production rate in the medium (production minus consumption) and D the dilution rate.

In the steady state, it is easily drawn from (3.7.1) that

$$\tilde{C}^{m}_{\mathrm{BUT}} = \frac{\Pi^{m}_{\mathrm{BUT}}}{D} \tag{3.7.2}$$

Assuming that $\Pi^{m}_{\mathrm{BUT}} = \text{constant}$, it is easily noted that the adjustment from (3.7.2) to the experimental points gives a satisfying result. The butyrate production rate can reasonably be considered as constant with $\Pi^{m}_{\mathrm{BUT}} = 0.012 \pm 0.001$ gBUT/(L h). The correlation coefficient is $r^2 = 0.87$. The result of the fitting appears in Fig. 3.34 (solid line).

Comments The difference between lactic production (that is only produced after a critical value of dilution rate) and the fermentation of butyrate (produced over the entire interval of D with constant rate) suggests that the fermentation of butyrate is done by some subpopulation of the bacterial floc. The constancy of the production rate seems to indicate that this subpopulation is relatively isolated from the external medium, perhaps in the core of the floc where it benefits from relatively stable conditions.

3.7.4.2 Lactate

The lactate production is singularly more complex to model. We used the same model as the one applied for the production of ethanol by *S. cerevisiae* in aerobic culture. This effect is known as the Crabtree effect and has been developed in great detail in the preceding section. So, only the most important results will be looked at again.

Figure 3.37 shows the principal metabolic flows considered in the model and can be described as follows.

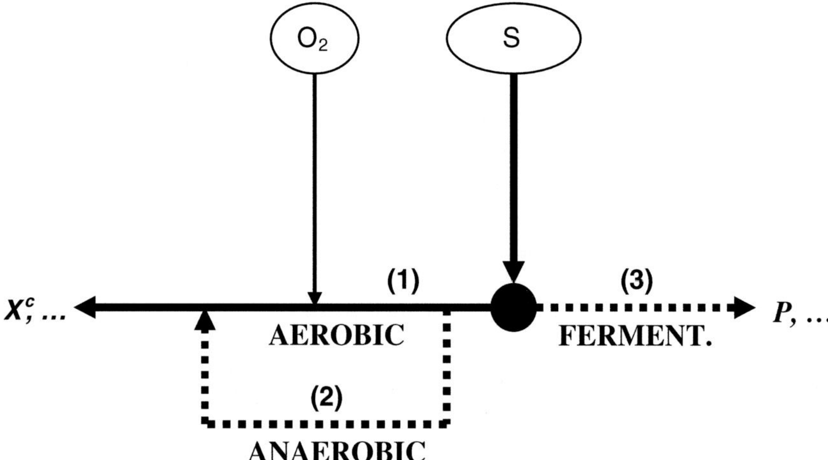

Fig. 3.37 Main specific metabolic fluxes taken into account by the model. Explanations: refer to the text

A substrate flow, S, penetrates into the system (the cell or the floc). At low dilution rates, the metabolic pathway (1) is the only one to be used to produce energy and biomass (as well as possible derived products such as water or CO_2). This pathway requires oxygen and so represents the aerobic pathway of the system. For a very precise critical dilution rate (D_c), the metabolic pathway (1) saturates and the flux of compounds derived from the substrate begin to borrow pathways (2) and (3); pathway (1) still remaining active. Pathway (2) contributes in an anaerobic manner to biosynthetic reactions. Pathway (3) is the fermentative pathway leading to the excretion of a product P (and possibly to derivatives such as CO_2). Above the critical dilution rate, since the aerobic metabolic pathway is saturated, the specific flow of oxygen remains constant. In this model, for example, the Crabtree effect in *S. cerevisiae* is not due to saturation of respiratory chain reactions (Sonnleitner and Käpelli 1986) but to the metabolic pathway (1) being saturated with derivatives from the substrate. Since the specific uptake flux of oxygen is proportional to the metabolic rate of the substrate in the pathway (1) (refer to Sect. 3.6), the saturation of this brings the specific oxygen uptake rate to a constant value. From this point of view, a sudden addition of substrate (a pulse) to the low dilution rates should lead to a sudden demand for oxygen (observable by a reduction in dissolved oxygen in the medium). On the contrary, for $D > D_c$, this phenomenon is not to be observed since the supplementary substrate cannot influence pathway (1) that is saturated.

Mathematically, the model is represented in the following manner. The general mass balance in the cell phase is (refer to (3.6.19)):

$$\Phi^0_{S,m}(c) = \left(q^c_s(h) + q^c_s(l)\right)X^c + D\tilde{C}^c_S \tag{3.7.3}$$

where $\Phi^0_{S,m}(c) = D\left(C^{m,E}_S - C^m_S\right)$ is the interphasic exchange flow of substrate S that leaves the matrix phase (the culture medium) to go to the cell phase: m $\rightarrow$ c. $\tilde{C}^{m,E}_S$ and $\tilde{C}^m_S$ are the substrate concentrations in the medium, respectively, at the inlet and in the reactor; $q^c_S(.)$ are specific rates of transport and metabolism (T/M) of S in the cell phase; X^c is the biomass. The model is limited to two T/M pathways.

The high-affinity pathway, $q^c_S(h)$, has the following form:

$$q^c_S(h) = \frac{V^0_S \tilde{C}^c_S}{K^*_S X^c + \tilde{C}^c_S}; \quad K^*_S \ll 1 \tag{3.7.4}$$

where $\tilde{C}^c_S$ is the intracellular concentration of substrate; V^0_S is the maximum specific T/M rate; $K^*_S X^c$ is a measure of the affinity (that depends on the biomass). The high-affinity pathway is called such because $K^*_S \ll 1$.

The low-affinity pathway, $q^c_S(l)$, is given by,

$$q^c_S(l) = k^*_0 \frac{\tilde{C}^c_S}{X^c} \tag{3.7.5}$$

where k^*_0 is a complex constant depending both on the maximum specific T/M rate, on the affinity and on the specific mass of the cell phase.

In the cellular phase, when there is excretion of a product that is derived from the substrate S (for example, by fermentation), the partial mass balance is written as,

$$\Phi^0_{S,m}(c) = \left(q^c_s(h) + \beta q^c_s(l)\right)X^c + D\tilde{C}^c_S \tag{3.7.6}$$

where β is the fraction of the low-affinity specific rate allocated to the pathway leading to excretion (pathway (3) in Fig. 3.36).

Naturally, the two balances (3.7.3) and (3.7.6) should be satisfied simultaneously. To do that, the partial balance is satisfied by resolving (3.7.6) with respect to $\tilde{C}^c_S$. Once this value is fixed (for a given value of D), the modification of the biomass resulting from the excretion can be calculated from (3.7.3):

$$X^{/c} = \frac{\Phi^0_{S,m}(c) - D\tilde{C}^c_S}{q^c_S(h) + q^c_S(l)} \tag{3.7.7}$$

It can be shown (refer to Sect. 3.6) that there is a critical dilution rate given by

$$D_c \approx V^0_S Y_{X^c,S} \tag{3.7.8}$$

for which a sudden transition occurs that causes a drop in biomass and the appearance of the excreted product in the medium ($Y_{X^c,S}$ is the initial yield coefficient to $D < D_c$).

Mass concentration of the excreted product can easily be calculated in the steady state by

$$\tilde{C}_P^m = \frac{q_P^c}{X/^c D} \tag{3.7.9}$$

where

$$q_P^c = g_S(v_i)(1-\beta)q_S^c(l) \tag{3.7.10}$$

$g_S(v_i)$ is a constant that depends on the stoichiometry of the transformation

$$v_S S \rightarrow v_P P + \ldots \tag{3.7.11}$$

3.7.5 Estimation of Subtrate in the Medium

In (3.7.3), X^c is given by $X^c = Y_{X^c,S}(\tilde{C}_S^{m,E} - \tilde{C}_S^m)$. The calculation of the profile X^c (D) necessitates knowledge of the values for $\tilde{C}_S^m(D)$ in the steady state. The concentration of the substrate in the medium must therefore be known for several values of D. The interpolation of the experimental points can be obtained by adjustment of the experimental points to an efficient arbitrary function. In this experiment, only the glucose is known. The regression was achieved using the function GLU = $a + bD + c(\ln D)^2 + d/D + e/D^{1.5}$. The correlation coefficient is $r^2 = 0.99$ and gives a good profile for glucose in the medium according to D (refer to Fig. 3.33). However, the flocs are cultivated on a complex medium, a mixture of glucose, peptone, and meat extract. The composition of such substrates is evidently not rigorously defined and it is difficult (if not impossible) to measure the residual substrate other than the glucose. It is a situation often encountered in natural media where the substrate is nearly always badly defined. In general, in these situations, there is recourse to global measures, such as COD (chemical oxygen demand) or TOC (total organic carbon), etc. These measurements also have errors because they do not necessarily represent the fraction that is really metabolizable by the cells. In the model described above, knowledge of the substrate interphasic exchange flux between the medium and the cell phase is indispensable. In fact, it represents the main constraint to which the cell system is subject. Now, we cannot quantify this quantity exactly without knowing the precise portion of peptone or meat extract consumed by the cells. So the problem does not seem to have a direct solution. To resolve this difficulty indirectly, the hypothesis is that a fraction of the complex nutriments (peptone and meat extract) can be measured as glucose equivalents.

The interphasic exchange flow for each nutrient is given by

$$\Phi_{i,m}^0(c) = D(\tilde{C}_i^{m,E} - \tilde{C}_i^m) \tag{3.7.12}$$

By summing over all the nutrients, it is obtained that

$$\Phi^0_{\mathrm{TOT},m}(c) = D\Big(\sum_i \tilde{C}_i^{m,E} - \sum_i \tilde{C}_i^m\Big) \tag{3.7.13}$$

which becomes

$$\Phi^0_{\mathrm{TOT},m}(c) = D(\tilde{C}^{m,E}_{\mathrm{TOT}} - \tilde{C}^m_{\mathrm{GLU}} - \tilde{C}^m_{\mathrm{ME}} - \tilde{C}^m_{\mathrm{Pe}}) \tag{3.7.14}$$

where $\tilde{C}^{m,E}_{\mathrm{TOT}}$ is the sum of the nutrients at the inlet of the chemostat; the indices are defined as follows: GLU: glucose; ME: meat extract; Pe: peptone. In (3.7.14), we are defining,

$$k(\tilde{C}^m_{\mathrm{ME}} + \tilde{C}^m_{\mathrm{Pe}}) \equiv \tilde{C}^m_{\mathrm{GLU/eq}} \tag{3.7.15}$$

where k is an equivalence factor between ME + Pe and the glucose. (This factor is obviously unknown a priori.) By omitting the index "/eq," (3.7.14) can then be put in the form

$$\Phi^0_{TOT,m}(c) = D(\tilde{C}^{m,E}_{\mathrm{TOT}} - (1+k)\tilde{C}^m_{\mathrm{GLU}}) \tag{3.7.16}$$

This notation is equivalent to the hypothetical transformation reaction

$$\nu_{\mathrm{ME}}\mathrm{ME} + \nu_{\mathrm{Pe}}\mathrm{Pe} \rightarrow \nu_{\mathrm{GLU}}\mathrm{GLU} \tag{3.7.17}$$

with the consequence that all the phenomena can be interpreted in terms of glucose equivalents. From a physiological point of view, a transformation such as (3.7.17) is very plausible, at least in theory. However, from the point of view of the model, this theory is very risky because it assumes that the structure of the model is invariant with respect to this transformation. The simulations appear in Fig. 3.33 (solid lines) and show, in an unexpected manner, that the profiles that were obtained are perfectly compatible with the experimental data. We obtained a value for the equivalent factor: $k = 1.5$ and was constant over the entire interval D. It is moreover easy to verify that $k = 1.5$ is a proper value when recalculating the initial yield coefficient (between $D = 0.10$ and $0.24\ \mathrm{h}^{-1}$). In this domain, $Y_{X^c,\mathrm{TOT}}$ is constant and at a value of about 0.423. If the value of k is correct over the whole of the domain, it should be observed, for the initial value and by assuming that the residual substrates are small, that in this domain

$$\frac{X^c}{\tilde{C}^{m,E}_{\mathrm{GLU}} + \tilde{C}^{m,E}_{\mathrm{ME}} + \tilde{C}^{m,E}_{\mathrm{Pe}}} \approx \frac{X^c}{\tilde{C}^{m,E}_{\mathrm{GLU}}(1+k)} \tag{3.7.18}$$

The right-hand side of (3.7.18) gives a value of 0.44, very close to 0.423 (left-hand side). The two values differ by only 4 %.

So, on the whole, the phenomenon of lactic fermentation is well represented by the model described by Eqs. (3.7.3–3.7.7) and (3.7.9) using the hypothesis (3.7.15) that makes it possible to represent the entire process in glucose equivalents.

3.7.6 Discussion

The first striking observation in this culture concerns the production modes that are completely different for the two fermentation products, butyrate and lactate. The first is produced in a constant way, independently of the growth rate of the flocs. The second, on the contrary, only appears in the medium for a growth rate that is greater than a critical value. The possibility that butyrate is secreted by a particular underpopulation, relatively well protected in the core of the floc has been mentioned. As far as lactate is concerned a model inspired by the long-term Crabtree effect in *S. cerevisiae* has been used and has given some very good results. Could it be concluded therefore that the observed phenomenon in the flocs culture is very comparable to the Crabtree effect? For this, the following points should be checked:

1. The appearance of lactate in the medium is well correlated with the fall in biomass.
2. Production of lactate is truly respiro-fermentative.

The first point is not immediate. In fact, examination in detail, of the experimental results Fig. 3.33 allows the assumption that the appearance of lactic acid occurs slightly earlier than the biomass drop. In fact, all takes place over an interval corresponding to 0.05 h^{-1} since the critical value of dilution rate that is obtained is of the order of 0.245 h^{-1}, whereas the appearance of lactate begins toward 0.2 h^{-1} and the fall in biomass toward 0.3 h^{-1}. In spite of the precautions taken, cumulated errors in the volumic flow rate and the volume of the reactor that are used to determine D are of about the order of 0.05, so that the results obtained are still within the margin of experimental error. Moreover, it is not reasonable to think that all bacteria of the floc produce lactic acid, so the delay in the biomass decrease can be explained by the fact that the nonlactic bacteria maintain the total biomass at a higher level near the critical point. To summarize, the correlation of the appearance of the fermented product and the fall in biomass is possible, but not absolutely confirmed by the experimental results. With respect to the respiro-fermentative character for the production of lactate, there is no direct evidence given that the oxygen fluxes have not been measured. However, it is known that at low dilution rates, a pulse of substrate leads to accumulation of oxygen, whereas beyond the critical rate, this fact does not occur. Should this phenomenon be associated with another subpopulation or due to an overall respiro-fermentative production of lactate? We cannot conclude at this stage of our research.

The hypothesis of a Crabtree effect due to a subpopulation of the consortium is defensible. In 1967 already, Muresian and Mustea (Mustea and Muresian 1967) presented a Crabtree effect in certain bacterial cultures (based on respirometry measurements). Wolfe describes at length the "acetate switch" under aerobic cultures of bacteria (Wolfe 2005). More recently, an extensive exometabolome analysis revealed a production of numerous subproducts in the culture medium and the authors refer to a bacterial Crabtree effect (Paczia et al. 2012).

However, it is also known that bacterial consortia included in biofilms can present particular properties that an isolate species does not possess (Adrian et al. 1998; Alexander 1999); for example, in the case considered, the transport phenomena for substrates could be widely different at the core of a floc or for isolated cells. In our model used, the transport phemomena have a crucial importance and the grouping in floc could therefore play an important role. To conclude on this subject, let us say that it is currently impossible to settle the question definitely. Nevertheless, the behavior of the culture greatly resembles, as regards certain global characteristics, a long-term Crabtree effect. However, this effect cannot yet be associated with one species or with one well-defined underpopulation. Practically, nevertheless, the model that has been applied really takes the main macroscopic characteristics into account as well. The diagram (3.7.5) can be applied by conferring a sort of physiological (or metabolic) unity to a bacterial floc, placing the level of description above than that of bacterial species or subpopulation. This conception of the floc as a functional entity is reinforced by the still unexplained fact (as far as we know) that the floc growth rate adjusts itself to the dilution rate of the chemostat, just as happens with isolated cells (mammals, yeasts, and bacteria). The question of the reproduction of a floc remains, nevertheless an open question.

We must, before concluding this chapter, point out that there is, however, one important difference between the situation outlined in *S. cerevisiae* and the consortium. In the first case, the critical dilution rate does not vary with the inlet concentration (Rieger et al. 1983), while it decreased with inlet glucose concentration for the consortia (Bensaid 2000). It is perhaps too early to present definitive conclusions, but these preliminary findings lead us to think that these results

(1) back up our hypothesis on the saturation of glucose metabolism (rather than the respiratory pathway);
(2) show the importance of transport phenomena;
(3) encourage us to think that the diffusion phenomenon of glucose toward the cell plays a role in the case of consortia but not in that of yeast. This last observation does not contradict at all, the fact that the same model is applied in both cases (refer to Chap. 2 and the interpretation in diffusional terms).

Finally, from a practical point of view, there are without a doubt perspectives on the level of industrial applications. The culture that grew showed that bacterial floc-forming consortium can be very stable (no sterilization necessary) and is likely to produce various fermentation products by a continuous process (here, the chemostat) and it did this in a way that could be reproduced and controlled. This observation has its importance. In fact, it is generally considered (Nielsen and Villadsen 1994) that continuous processes can be industrially more competitive but that the contamination problems limit their intensive practice. Production of certain derivatives of interest to industry by strong consortia could represent a new and fruitful paradigm in this field.

References

Adrian L., Manz W., Szewzyk U. and Görisch H. (1998) Physiological characterization of a bacterial consortium reductively dechlorinating 1,2,3- and 1,2,4-trichlorobenzene Appl. Envir. Microbiol. **64**:496-503.

Alexander, M. (1999) Biodegradation and Bioremediation (2[d] edition) Academic Press, London, UK.

Alim S. & Ring, K. (1976) Regulation of amino acid transport in growing cells of *Streptomyces hydrogenans*. II. Correlation between transport capacity and growth rate in chemostat cultures. Arch. Microbiol. **111**: 105-110.

Aon J.C. and Cortassa S. (2001) Involvement of nitrogen metabolism in the triggering of ethanol fermentation in aerobic chemostat cultures of *Saxxharomyces cerevisiae* Met. Eng. **3**:250-264.

Abbott, A.J. & Nelsestuen, G.L. (1988) The collisional limit: an important consideration for membrane-associated enzymes ad receptors. *FASEB J.* **2**, 2858-2866.

Baldwin, W.W. & Kubitschek, H.E. (1984) Buoyant density variation during the cell cycle of *Saccharomyces cerevisiae*. J. Bacteriol. **158,** 701-704.

Barford J.P. and Hall R.J. (1979) An examination of the Crabtree effect in *Saccharomyces cerevisiae*: The role of the respiratory adaptation. J. Gen. Microbiol. **114**: 267-275.

Barford JP, Jeffrey PM and and Hall RJ (1981) The Crabtree effect in *Saccharomyces cerevisiae* – primary control mechanism or transient In; Advances in Biotechnology 1 – M. Moo-Young Ed. Pergamon Press, Englewwod Cliffs, New Jersey pp. 255-260.

Bellgardt, K.H. (1991) Cells Model. In: *Biotechnology* (Rehm, H.J. & Reed, G. with Pühler, A & Stadler, eds) *Vol. 4* (K. Schügerl, ed) pp 267-298. Weinheim: VCH.

Bellgardt K.H. (2000a) Baker's yeast production (p. 277-320) In: Bioreaction Engineering; K Schügerl and KH Bellgardt (Eds) Springer Verlag, Berlin, Heidelberg.

Bellgardt K.H. (2000b) Bioprocess models. In: Bioreaction Engineering; K Schügerl and KH Bellgardt (Eds) Springer Verlag, Berlin, Heidelberg.

Bensaïd, A (2000) Comportement complexe d'un consortium bactérien dans un chemostat. Ph. D. Thesis, Faculté des Sciences, Université Libre de Bruxelles, Brussels.

Bensaid A., Thierie J. and Penninckx M. (2000) The use of tetrazolium salt XTT for the estimation of biological activity of activated sludge cultivated under steady-state and transient regimes. J. microbiol. methods **40**:255-263.

Blake-Coleman B.C. (1993) Approach to physical measurements in biotechnology Academic Press, London - ISBN 0-12-103610-3.

Bond D.R. and Russell J.B. (1998) Relationship between Intracellular Phosphate, Proton Motive Force, and Rate of Nongrowth Energy Dissipation (Energy Spilling) in *Streptococcus bovis* JB1. Appl. Environ. Microbiol. **64**(3): 976-981.

Bond D.R. and Russell J.B. (2000) Protonmotive force regulates the membrane conductance of *Streptococcus bovis* in a non-Ohmic fashion. Microbiol. **14**: 687-694.

Brown D. and Rothery P. (1993) Models in biology. Mathematics, Statistics and Computing. Wiley & Sons, Chichester, England.

Button D.K. and Kinney P.J. (1978) Unidiretional flux determination by isotope relaxation methods: theory Contin. Cultiv. Microorganisms, Proc. 7th Symp., Prague, July 10-14, Prague (1980).

Button D.K. (1991) Biochemical basis for whole-cell uptake kinetics: Specific activity, Oligotrphic capacity, and the meaning of the Michaelis constant Appl. Environ. Microbiol. **57** (7):2033-2038.

Carlson M (1987) Minireview: Regulation of sugar utilization in *Saccharomyces* species. J. Bacteriol. **169**: 4873-4877.

Chandrasekhar S. (1970) Hydrodynamic and hydromagnetic stability. Oxford University Press, London. (Third Edition).

Chesbro W., Evans T. and Eifert R. (1979) Very slow growth of *Escherichia coli*. J. Bacteriol. **139**: 625-638.

Cole J.R., Olsson C.L.; Hershey JWB; Grunberg-Manago M & Nomura M (1987) Feedback regulation of rRNA synthesis in Escherichia coli: Requirement for initiation factor IF2. J. Mol. Biol. 198:383-392.

Contois, D.E. (1959) Kinetics of bacterial growth: relationship between population density and specific growth rate of continuous cultures. J. Gen. Microbiol. **21:** 40-50.

Cortassa S and Aon MA (1997) Distributed control of the glycolytic flux in wild-type cells and catabolite repression mutants of *Saccharomyces cerevisiae* growing in carbon-limited chemostat cultures Enz. Microb. Technol. **21**: 596-602.

Cortassa S and Aon MA (1998) The onset of fermentative metabolism in continuous cultures depends on the catabolite repression properties of *Saccharomyces cerevisiae* Enz. Microb. Technol. **22**: 705-712.

Cortassa S., Aon M.A., Iglesias A.A. and Lloyd D. (2002) An introduction to metabolic and cellular engineering. World Scientific publishing, Singapore.

Costerton J.W. and Lappin-Scott H.M. (1989) Behavior of bacteria in biofilms ASM News **55**(12): 650-6540.

Costerton J.W., Lewandowski Z., Caldwell D.E., Korber D.R. and Lappin-Scott H.M. (1995) Microbial biofilms Annu. Rev. Microbiol. **49**:711-745.

Coulson, J.M. & Richardson, J.F. (1987) Chemical engineering (Second edition.) Pergamon Press, UK.

Crabtree HG (1929) Observations on the carbohydrate metabolism of tumours. Biochemical Journal **23**: 536-545.

D'Amore T, Panchal CJ and Stemart GG (1988) Intracellular ethanol accumulation in *Saccharomyces cerevisiae* during fermentation Appl. Envir. Microbiol. **54**(1): 110-114.

Decho, A.W. (1999) Chemical communication within microbial biofilms: chemotaxis and quorum sensing in bacterial cells. In: *Microbial extracellular polymeric substances* (Wingender, J., Neu, T.R. & Flemming, H.-C., eds) pp 155-165. Berlin: Spinger-Verlag.

de Deken R.H. (1966) The Crabtree Effect: A Regulatory System in Yeast. J. Gen. Microbiol. **44**:149–156.

de Hollander J.A. (1993) Kinetics if microbial products formation and its consequences for the optimization of fermentation processes. Antonie van Leeuwenhoek, **63**: 375-381.

Demain A.L. and Solomon N.A. (1986) Manual of industrial microbiology and biotechnology. American Society for Microbiology, Washington, D.C.

Doran P.M. (1995) Bioprocess Engineering Principles Academic Press, London, UK.

Doran P.M. (1995) Bioprocess Engineering Principles. Academic Press Limited, London.

Galadzhiev M.A. (1932) On the problem of protozoa immortality. Proc. USSR Acad. Sci. **9**:1269-1300.

Guijarro JM and Lagunas R (1984) *Saccharomyces cerevisiae* does not accumulate ethanol against concentration gradient J. Bacteriol. **60**(3): 874-878.

Guiseppin ML and van Riel NAW (2000) Metabolic modeling of *Saccharomyces cerevisiae* using the optimal control of homeostasis: A cybernetic model definition. Met. Eng. **2**:14-33; doi: ID mben.1999.0134.

Glansdorf P. and Prigogine I. (1971) Structure, stabilité et fluctuations. Masson et Cie., Paris.

Glazer N. and Nikaido H. (1995) Microbial Biotechnology - Fundamentals of applied microbiology W.H. Freeman & C°, USA.

Golle H.A. (1953) Theoretical considerations of continuous culture systems. Agric. Fed. Chem. **1**:1789-1798.

Herbert D. (1961) The chemical composition of microorganisms as a function of their environment. 11th Symp. Soc. Gen. Microbiol. **11:** 391-416 Cambridge University Press.

Ingraham F.C., Maaloe O. and Neidhart F.C. (1983) Growth of the bacterial cell. Sinauer Associates, Inc., Sunderland, USA.

Kubitschek, H.E., Baldwin, W.W. & Graetzer, R. (1983) Buoyant density constancy during the cell cycle of *Escherichia coli. J. Bacteriol.***155,** 1027-1032.

Kubitschek, H.E., Baldwin, W.W., Schroeter, S.J. & Graetzer, R. (1984) Independance of buoyant density and growth rate in *Escherichia coli. J. Bacteriol.***158,** 296-299.

Kuenen J.G. (1979) Growth yields and "maintenance energy requirement" in *Thiobacillus* species under energy limitation. Arch. Microbiol., **122**:183-188.

Kuznetsov Y.A. (1995) Elements of applied bifurcation theory. Springer-Verlag, New York.

Kwong S.C.W. and Rao G. (1994) Metabolic monitoring using the rate of change of NAD(P)H fluorescence. Biotechnol. Bioeng. **44**:453-459.

Lagunas R, Dominguez C, Busturia A and Sáez MJ (1982) Mechanisms of appearance of the Pasteur effect in *Saccharomyces cerevisiae*: Inactivation of sugar transport systems. J. Bacteriol. **152**(1): 19-25.

Lagunas R (1992) Sugar transport in *Saccharomyces cerevisiae* FEMS Microbiol. Rev. **104**:229-242.

Lagunas R (1993) Sugar transport in Saccharomyces cerevisiae. FEMS Microbiol Rev **10**:229–242.

Liu Y. and Chen G.H. (1997) Model of energy uncoupling for substrate-sufficient culture. Biotechnol. Bioeng. **55**(3): 571-576.

Marr A.G. (1991) Growth Rate of *Escherichia coli.* Microbiol. Rev. **55(2)**: 316-333.

McClung L.S. (1949) Recent developments in microbiological techniques. Rev. Microbiol. **3**:395-422.

Monod J. (1942) Recherches sur la croissance des cultures bactériennes. Hermann, Paris.

Monod J. (1949) The growth of bacterial cultures. Ann. Rev. Microbiol. **3** : 371-394.

Monod J. (1950) La technique de culture continue : théorie et applications. Ann. Instit. Pasteur. **79** : 390-410.

Mustea I. and Muresian T. (1967) Crabtree effect in some bacterial cultures. Cancer **20**:1499–14501.

Neidhardt F.C., Ingraham J.L. and Schaechter M. (1995) Physiologie de la cellule bactérienne - Une approche moléculaire. Masson, Paris.

Neijssel O.M. and Tempest D.W. (1975) The regulation of carbohydrate metabolism in *Klebsiella aerogenes* NCTC 418 organisms growing in chemostat culture. Arch. Microbiol. **106**: 251-258.

Neijssel O.M and Tempest D.W. (1976) The role of energy-spilling reactions in the growth of Klebsiella aerogenes NCTC 418 in aerobic chemostat culture. Arch Microbiol. **110**:305–11.

Nicolis, G. and Prigogine I. (1977) Self-organization in nonequilibrium systels. From dissipative structures to order through fluctuations. John Wiley and Sons, New York, USA.

Nielsen J. and Villadsen J. (1994) Bioreaction engineering principles. Plenum Press, New York, USA.

Novick A. and Szilard L. (1950) Description of the chemostat. Science **112**:715-716.

Paczia N., Nilgen A., Lehmann T., Gätgens J., Wiechert W. and Noack S. (2012) Extensive exometabolome analysis reveals extended overflow metabolism in various microorganisms. Microbial Cell Factories **11**:122–136.

Panikov N.S. (1995) Microbial growth kinetics. Chapman & Hall, London.

Pirt, S.J. (1965) The maintenance energy of bacteria in growing cultures. Proc. R. Soc. London Serie B Biol. Sci. **163:** 224-231.

Pirt, S.J. (1982) Maintenance energy: a general model for energy-limited and energy-sufficient growth. Arch. Microbiol. **133,** 300-302.

Postma, E., Scheffers, W.A. & van Dijken, J.P. (1989) Kinetics of growth and sugar transport in glucose-limited chemostat cultures of *Saccharomyces cerevisiae* CBS 8066. *Yeast.***5,** 159-165.

Prigogine, I. (1968) Introduction à la thermodynamique des processus irréversibles. Monographies Dunod, Dunod, Paris.

Rieger M, Käpelli O and Fiechter A (1983) The role of limited respiration in the complete oxidation of glucose by *Saccharomyces cerevisiae* J. Gen. Microbiol. **129**: 653-661.

Robertson, B.R, and Button D.K. (1979) Phosphate-limited continuous culture of *Rhodotorula rubra*: kinetics of transport, leakage, and growth. J. Bacteriol. **138**:884–895.

Roels, J.A. (1983) *Energetics and kinetics in biotechnology.* Elsevier Biomedical Press; The Netherlands.

Roques, H., Yue, S., Saipanich, S. & Capdeville, B. (1982) Faut-il abandonner le modèle de Monod pour la modélisation des processus de dépollution biologique? *Water Res.* **16,** 839-847.

Rosenberger R.F. and Kogut M. (1958) The influence of growth rate and aeration on the respiratory and cytochrome system of a fluorescent Pseudomonad grown in continuous culture. J. Gen. Microbiol. 19:228-243.

Russell J.B. and Cook G.M. (1995) Energetics of Bacterial Growth: Balance of Anabolic and Catabolic Reactions. Microbiol. Rev. **59**: 48-62.

Schechter E. (1997) Biochimie et biophysique des membranes – Aspects structuraux et fonctionnels Masson, Paris.

Schügerl, K. & Bellgardt, K.H., eds (2000) *Bioreaction engineering.* Springer-Verlag, Berlin.

Segel L.A. (1984) Modeling dynamic phenomena in molecular and cellular biology. Cambridge University Press, USA.

Senn H.U., Lendenmann U., Snozzi M., Hammer G. and Egli T. (1994) The growth of *E. coli* in glucose-limited chemostat cultures: a reexamination of kinetics. Biochim. Biophys. Acta. **1201**:424-436.

Shapiro J.A. (1991) Multicellular behavior of bacteria. ASM News, **57**: 247-253.

Skellam J.G. (1973) The formulation and interpretation of mathematical models of diffusionary processes in population biology. In Bartlett M.S. & Hiorns R.W (Eds) *The mathematical theory of the dynamics of biological populations.* Academic Press, New York, 63-85.

Sonnleitner B. and Käpelli O. (1986) Growth of *Saccharomyces cerevisiae* is controlled by its limited respiratory capacity: Formulation and verification of a hypothesis. Biotechnol. Bioeng. **28**: 927-937.

Streekstra M., Teixeira de Mattos J., Neijssel O.M. and Tempest D.W. (1987) Overflow metabolism during anaerobic growth of *Klebsiella aerogenes* NCTC 418 on glycerol and dihydroxyacetone in chemostat culture. Arch. Microbiol. **147**: 268-278.

Tempest D.W., Hunter J.R. and Sykes J. (1965) Magnesium limited growth of *Aerobacter aerogenes* in a chemostat. J. Gen. Microbiol. 39:355-366.

Tempest D.W. and Neijssel O.M. (1984) The status of Y_{ATP} and maintenance energy as biologically interpretable phenomena. Ann. Rev. Microbiol. **38**:459-486.

Thierie J., Bensaid A. and Penninckx M. (1996) Transient phenomena in cultivated activated sludge observed after increases in chemostat dilution rate. - Arch. Soc. Belg. Biochem., **B38.**

Thierie J. (1997) Why does bacterial composition changes with the chemostat dilution rate ? Biotechnol. Techniques **11**:625-629.

Thierie J., Bensaid A. & Penninckx M. (1999) Robustness, coherence and complex behaviors of a bacterial consortium from an activated sludge cultivated in a chemostat. Med. Fac. Landbouw, Univ. Gent, **64/5a**:205-210 / Thirteenth Forum for Applied Biotechnology, Gent, 1999.

Thierie, J. (2000) Cellular cycling of substrate as a possible cryptic way for energy spilling in suspended cellular continuous cultures. *Biotechnol. Let.***22:** 1143-1149.

Thierie, J (2004a) Modeling threshold phenomena, metabolic pathways switches and signals in chemostat cultivated cells: the Crabtree effect in *Saccharomyces cerevisiae.* J. Theor. Biol. **226**:483-501.

Thierie, J (2004b) Modeling threshold phenomena, metabolic pathways switches and signals in chemostat cultivated cells: Theoretical results. J. Theor. Biol. (additional data on line).

Thierie J. & Penninckx M. (2004) Possible occurrence of a Crabtree effect in the production of lactic and butyric acids by a floc forming bacterial consortium. Current Microbiology **48** (3) :224-229 (march 2004).

Torres N.V., Voit E.O., Glez-Alcòn C. and Rodriguez F. (1997) An indirect Optimization Method for Biochemical Systems: Description of Method and Application to the Maximization of the Rate of Ethanol, Glycerol, and Carbohydrate Production in Saccharomyces cerevisiae. Biotechnol. and Bioeng. 55(5):758-772.

Tsai S.P. and Lee Y.H. (1990) A model for Energy-Sufficient Growth. Biotechnol. Bioeng. **35**:138-145.

Van Urk H., Mak P.R., Scheffers W.A. and Van Dijken J.P. (1988) Metabolic response of *Saccharomyces cerevisiae* CBS 8066 and *Candida utilis* CBS 621 upon transition from glucose limitation to glucose excess. Yeast **4**:283-291.

Van Urk H, Postma E, Scheffers WA and Van Dijken JP (1989) Glucose transport in Crabtree-positive and Crabtree-negative yeasts. J. Gen. Microbiol. **135**: 2399-2406.

Varma, A. & Palsson, B.O. (1993a) Stoechiometric interpretation of *Escherichia coli* glucose catabolism under various oxygenation rates. *Appl. Env. Microbio.l***59:** 2465-2473.

Varma, A. & Palsson, B.O. (1993b) Metabolic capabilities of *Escherichia coli* : II. Optimal growth patterns. *J. Theor. Biol.***165:** 503-522.

Varma A. and Palsson B.O. (1994) Stoechiometric flux balance models quantitatively predicts growth and metabolic by-products secretion in wild-type *Escherichia coli* W3110 Appl. Envir. Microbiol. **60**: 3724-3731.

Varner J. and Ramkrishna D. (1999) Metabolic engineering from a cybernetic perspective. 1. Theoretical preliminaries. Biotechnol. Prog. **15**:407-425.

von Meyenburg HK (1969) Katabolit-Repression und der Sprossung zyklus von *Saccharomyces cerevisiae* Ph. D. thesis, Eidgenössische Technsiche Hochshulle, Zürich.

Villermaux J. (1982) Génie de la réaction chimique. Conception et fonctionnement des réacteurs. Technique et Documentation (Lavoisier), Paris.

Walker G.M. (1998) Yeast – Physiology and Biotechnology John Wiley & Sons Ltd, Chichester, England.

Walsh MC, Smits HP, Scholte M and van Dam K (1994) Affinity of glucose transport in *Saccharomyces cerevisiae* is modulated during growth on glucose J Bacteriol **176**: 953-958.

Walsh MC, Scholte M, Valkier J, Smits P and van Dam K (1996) Glucose sensing and signaling properties in *Saccharomyces cerevisiae* require the presence of at least two members of the glucose transporter family J Bacteriol **178**: 2593-2597.

Waltman, Paul (1983) Competition models in population biology. Soc. Industrial and Applied Math., CBMS-NSF, n°45, Philadelphia.

Westerhoff H.V., Lokema J.S., Roel O. and Hellingwerf K.J. (1982) Thermodynamics of growth Non-equilibrium thermodynamics of bacterial growth. The phenomenological and the mosaic approach. Biochim. Biophys. Acta **683**:181-220.

Weusthuis, R.A., Pronk, J.T., van den Broek, P.J.A. & van Dijken, J.P. (1994) Chemostat cultivation as a tool for studies on sugar transport in yeasts. *Microb. Rev.***58,** 616-630.

Wojtczak L (1996) The Crabtree effect: a new look to the old problem Acta. Biochim. Pol. **43**(2): 361-368.

Wolfe AJ (2005) The acetate switch. Microbiol Mol Biol Rev. **69**:12–50.

Woodruff L.L. and Baitsell G.A. (1911a) The reproduction of *Paramecium aurelia* in a "constant" culture medium of beef extract. J. Exp. Zool. **11**: 135-142.

Woodruff L.L. and Baitsell G.A. (1911b) Rhythms in the reproductive activity of infusoria. J. Exp. Zool. **11**: 339-359.

Chapter 4
General Discussion

Abstract In this discussion, we try to identify some key ideas and to make a number of syntheses of the preceding results. Some important points are resumed and reexamined from another point of view, either theoretically or as examples. As final conclusion, it states that despite the excellent results obtained by studying the biphasic chemostat the polyphasic dispersed systems (PDS) theory is still largely in a state of incompleteness. The hope is expressed that some will be tempted to continue building this theory which has not until now revealed its full potential.

4.1 About the Theory

The principal objectives of this book were mainly to examine the possibility of describing complex polyphasic systems at different levels of description by using a unified theory. Theoretically, the approach can be applied to chemical, physical, or biological systems as long as they fulfill certain conditions of which the premises are given in Sects. 2.1 and 2.2.

We have chosen to study cultures of microorganisms in a particular open bioreactor namely the chemostat. The complexity of such a system is enormous, contrary perhaps to conventional wisdom. It is in fact made up of cells (or groups of cells), forming open subsystems, far from equilibrium, constantly in division and eventually subjected to forced inputs. These open subsystems are themselves parts of an open system (the chemostat as a whole). Finally, the majority of different phases that are considered are only maintained thanks to energy being supplied to the system (such as agitation, for example) and so constitute in a thermodynamic sense, an unstable system.

The question of knowing if there exists a rational, systematic, and nonempirical way of describing such systems, retaining a manageable formalism, is not necessarily straightforward. That is, what was attempted in Chap. 2. The approach consisted of writing the mass balances (or matter balances) for each compound at the core of micelles (a phase fraction) and by summation raising the degree of level

J. Thierie, *Introduction to Polyphasic Dispersed Systems Theory*,
DOI 10.1007/978-3-319-27853-7_4

of description by then obtaining the balance at phase level. Again by summation, the level of the whole system is finally reached. Finally, rather than keeping a structured description implying the description of each micelle we opted for a nonstructured representation at the micellian phase level, assuming that the notion of mean value could be used for the state variables significant for the description of the system.

For practical reasons, that cause experimental measurements to be obtained mainly at the phase and/or at the whole system level, the use of pseudo-homogeneous concentrations has been introduced and emphasized. These concentrations are applicable to systems that are in reality heterogenous. Obviously, the intimate dispersion of phases between them makes this pseudo-homogenous approach possible. It is the key characteristic of polyphasic dispersed systems (PDS).

By means of the prerequisites used above, we have been able to:

1. Derive a general pseudo-homogeneous mass balance for each compound in each phase of the system (cf. Eq. 2.4.6);
2. Bring to light in a systematic manner the interphasic exchange fluxes (cf. Fig. 2.6 and relationships (Eq. 2.4.3) and following);
3. Derive a relationship that links the concentrations describing two levels of description (local and extensive; cf. Eq. 2.3.17).

The derivation of the pseudo-homogenous balance (Point 1 above) required some effort and notably that particular attention be paid to perfecting a theory that was sufficiently practical and readable.

Concern for generalization required the introduction of terms that could not necessarily be interpreted immediately (refer notably to the variational term relating to the number of micelles, a subject only touched on in Appendix A.5). Interpretation seeking has moreover led to very promising considerations concerning the link between the number of cells, biomass, and doubling time, unfortunately still too incomplete to figure in this discussion). Finally, attempts were made to find the global results, obtained by other methods of representation, by adding together all products and/or all phases. This stage of general checking is illustrated in Appendix A.3, where it appears that the PDS approach makes it possible to find classical expressions such as the law of evolution of biomass while authorizing more refined interpretations.

Obtaining the pseudo-homogenous balance was obviously an obligatory step and one that showed itself to be fruitful, but it was points 2 and 3 above that proved to be more heuristic.

First, the division of a single system into subsystems obviously leaves the mass balance invariant. If the subsystems are not isolated one from another (and this would be a trivial case), the only way to keep the general matter balance invariant is to introduce exchange flows between subsystems. This is just what was done by defining the interphasic flows (point 2 above). It is the introduction of this type of flow (that is cancelled at the level of the whole system) that made it possible to develop the mechanism for dissipation of cell energy in Sect. 3.4.

Even in an apparatus as simple as the biphasic chemostat, a substrate recirculation flux can be conceived and quantified only by bringing to light the interphasic exchange flux. This recirculation flux implies a consumption of energy (it is not a question of an equilibrium reaction between intra- and extracellular substrate) and so constitutes an extremely simple mechanism for energy dissipation. Today, we thought that the reasoning begun on this badly resolved problem of the question of maintenance deserves to be reexamined within the wider perspective of the dissipation of cell energy. Representation in terms of PDS will perhaps establish better a thermodynamic interpretation of intracellular processes (that are far from being resolved; Blumenfeld and Tikhonov 1994).

Finally, point 3, that made it possible to establish a relationship between the expression of concentrations of the same compound at two different levels of description is perhaps the one that brought the most unexpected results in the expression of transport/metabolism kinetics obtained in Sects. 3.5 and 3.5.6. This particular point in fact cases the affinity for the substrate to appear as independent of the biomass in an extremely natural way and confirms with great simplicity difficult theoretical results obtained by other approaches or takes into account experimental results obtained and confirmed sometimes long ago.

To summarize, it is thought that it has been clearly shown in a direct or indirect manner that the theoretical results obtained in Chap. 2 are correct and that the PDS approach can well be used to produce a representation of systems of several levels of description.

Moreover, Chap. 3, that deal with applications of results obtained in Chap. 2, validly illustrates the heuristic perspective of the initiative and the wealth of interpretations that can stem from this.

However, in all honesty, the bearing on the success of the applications should be put in perspective because of the narrowness of its field of application. It was said before that the results of Chap. 2 are very general, but all applications from Chap. 2 concern only the stationary states of the biphasic chemostat. Consequently, the number of situations that are yet unexplored is enormous and constitutes, in a way, the perspectives of checking the theoretical results. It is illusory to want to draw up an exhaustive list of them, but it is mainly regretted that there was neither the time not the means:

- to explore other systems (open or otherwise) than the chemostat;
- to approach transient states of the biphasic chemostat;
- to approach situations in the chemostat with a higher number of phases, and in particular, with several different cell phases. This last case is particularly interesting because it constitutes a new theory of the dynamics of populations;
- to be able to confront the theory with systems of high cell density (another interesting characteristic for the PDS);
- …

It is easily ascertained that the PDS approach is in its infancy and that the number of interesting situations that could be looked at exceeds by far the number of cases studied. Each new situation is a challenge to the theory and it is much too soon to evaluate the extent of the field of application of this method of representation. Nevertheless, the results obtained to date are encouraging and from the point of view taken justify making an effort.

4.2 On the Subject of the Applications

So from the beginning, the applications only aimed to illustrate the interest of the PDS theoretical approach.

Two examples are cursorily approached at the beginning of Chap. 3 from this perspective. Here is illustration of demonstration [by using the relationship (Eq. 3.3.16)] that the specific synthesis rate of cytochromes from a *Pseudomonas* KB1 is independent of the dilution rate both in a medium with limited succinate and in a culture with limited air (Fig. 3.6). Obtaining such results (unconsidered by the authors) has been just about automatic if analyses of simple cases of the biphasic chemostat are used. These were developed in Sect. 4.3. On the other hand the methodology for the dosage of cytochromes, based on a difference in optical densities of 418 and 500 nm can perhaps be criticized and judged inadequate. The example serves as an illustration of the use of the PDS, but perhaps has no greatly supported biological signification.

Conversely, example 2 is based on a methodology that seems irreproachable and the approach used led to interpretations that did not necessarily coincide with those of the authors. For the record, the essentials of the authors' (Tempest et al. 1965) were that the synthesis of RNA from *Aerobacter aerogenes* NTC 418 is different in a culture medium with limited carbon (C-lim) or in a medium with limited magnesium (Mg-lim), which would imply, according to them, that there is regulation of the synthesis of RNA by magnesium as well as a variation in the activity of the ribosomes, linked to the Mg^{2+} cation.

It has been shown (Fig. 3.19) that the ratio of specific production rates for the RNAs and the proteins varied with dilution rate but was very independent of the medium. This observation, obviously does not encourage the theory of ribosomal activity governed by magnesium. How can the diversity of the interpretations be explained? We thought that it is a question of incorrect usage of the partial pseudo-homogeneous concentrations (E-concentrations) and of the reaction concentrations (or micellian, R-concentrations).

Figure 4.1 reproduces the essential information from which the authors draw their main conclusions (the influence of the medium and therefore of Mg^{2+}). It is observed that in fact the two curves differ greatly according to culture medium used. However, note that the units of the coordinate axes are mass per unit of useful volume (volume of culture), that is to say pseudo-homogeneous concentrations (E-concentrations), that would be noted as $\tilde{C}^{c}_{\mathrm{ARN}}$. All the data is not available (including the cell water content) that is needed to calculate the intra-micellian concentrations (or R-concentrations), but by conceding that the water content and

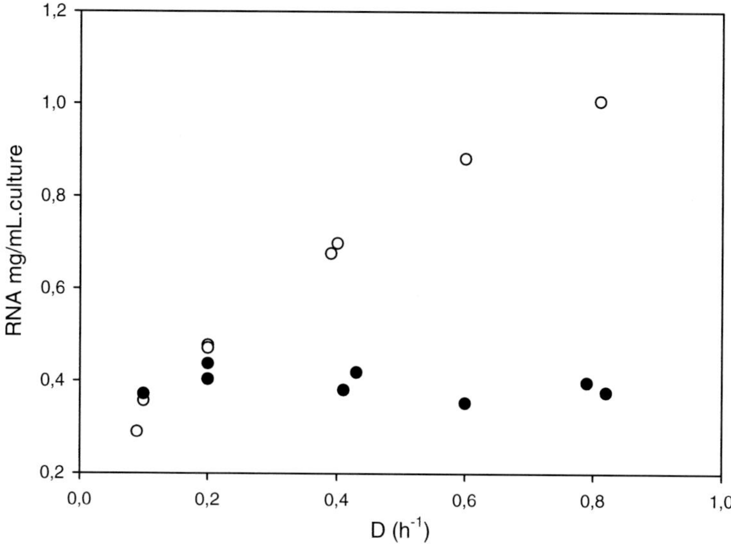

Fig. 4.1 Variation in RNA expressed in E-concentrations *versus D*. RNA concentration of RNA is expressed according to the dilution rate and one medium compared with another: *filled circle* Mg-lim; *open circle* C-lim. The profiles are different in the two media when they are expressed as E-concentrations (An anomalous point has been eliminated from the original)

the density of the cells does not vary too much from one medium to another, the following approximation can be conceded,

$$C^c_{\mathrm{ARN}} \cong k \frac{\tilde{C}^c_{\mathrm{ARN}}}{X^c} \tag{4.1}$$

where k is a constant that depends on the water content and on the cell density.

Figure 4.2 shows the RNA pool (mass of RNA per dry weight unit), a quantity proportional to the R-concentration, if the validity of (Eq. 4.1) is conceded.

Just as in the case of specific production rate, there are no longer any significant differences between the R-concentrations observed in the two media. So the new conclusion is that the influence of the medium is not significant as regards the intracellular concentration of RNA. Biologically speaking, it is this result in Fig. 4.2. that is relevant and not the interpretation based on Fig. 4.1. Of course, if it was a question of producing and extracting cellular RNA, Fig. 4.2. would clearly indicate that the C-lim medium is more effective that the medium with limited magnesium. The Mg-lim medium is therefore a medium that is less well adapted to the growth of *Aerobacter aerogenes* but the efficiency of the RNA/proteins system does not significantly depend on the medium.

In conclusion on this subject, it could be said that this approach not only does not validate the hypothesis of the authors of the change in ribosomal activity, according to medium (an eventuality that they, moreover, foresaw p. 365) but on the contrary, we bring to light a remarkable adaptation to medium from the perspective of the ratio

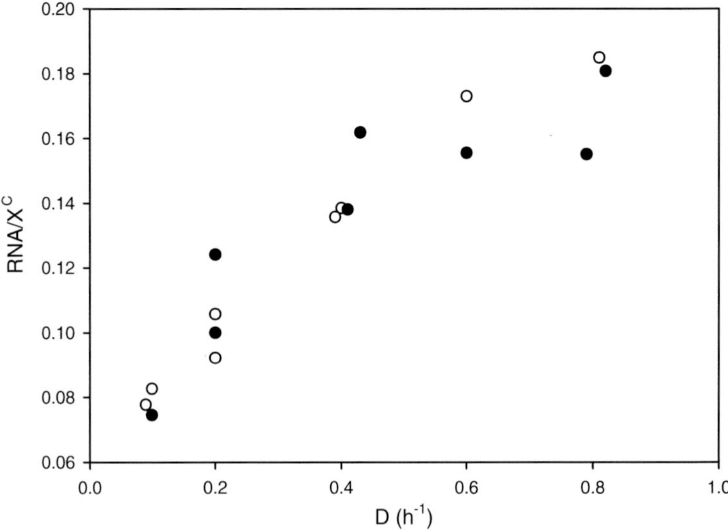

Fig. 4.2 Variation in RNA Expressed in R-concentrations *versus* D. RNA concentration is expressed as a function of the dilution rate and the two media compared: *filled circle* Mg-lim; *open circle* C-lim. The profiles cease to be significantly different between the two media when they are expressed in R-concentrations. The relationship 4.1 was used with $k = 1$ to estimate a quantity proportional to the R-concentration

of specific RNAs and proteins synthesis rates. The multiple and important roles of magnesium in bacterial growth have been well known for a long time (Mandelstam et al. 1982), but transport and homeostasis mechanisms of this cation have only been studied more recently (Smith et al. 1993; Smith and Maguire 1995; Gardner 2003). (This might explain why Tempest and his colleagues did not think it useful to dose Mg^{2+} into the media and they were able to consider that the cation could circulate freely from the culture medium to the intracellular medium.)

Regarding the applications that are the theme of Sects. 3.4–3.7, a good part of the comments were made in the body of the same text. What we would try here is a sort of synthesis of the most general observations that seem to stem from the two major examples that have been considered, namely the cell energy dissipation (Sect. 3.4) and the respirofermentative metabolic change (Sect. 3.6).

It is quite singular to note a posteriori, that the explanation of the case of dissipation of energy and that of aerobic fermentation to produce EtOH requires the same condition to re-balance the disturbed mass balances. In the two cases, the readjustment of the balances is obtained by modifying the biomass through an expression of the type,

$$X^c \rightarrow X^{/c} \quad \text{such as} \quad X^{/c} = \Psi(\Phi_S^0,\ q_S^c, \ldots) \tag{4.2}$$

(Refer to Eq. 3.4.12 for dissipation and Eq. 3.6.23 for the Crabtree Effect.)
From the two cases, it is obtained that

$$X^{/c} < X^c \tag{4.3a}$$

and

$$Y^{/}_{X^{/c},S} < Y_{X^c,S} \tag{4.3b}$$

In reality, in Sect. 3.4, without approximation (Eq. 3.4.9), it would be obtained (refer to Eq. 4.2):

$$\Psi_1 = \frac{\Phi_S^0 - D\tilde{C}_S^c}{q_S^c + \Delta q_S^c} \tag{4.4a}$$

(where $\Phi_S^0 \equiv \Phi_{\text{obs}}^0$; refer to Eqs. 3.4.5 and 3.6.4) and in Sect. 3.6, it is that:

$$\Psi_2 = \frac{\Phi_S^0 - D\tilde{C}_S^c}{q_S^c(h) + q_S^c(l)} \tag{4.4b}$$

Indeed, in both cases there is an excretion of the substrate, whether without modification (as in the case of dissipation) or after moderate oxidation (*GLU* → *EtOH*) in the case of the Crabtree Effect. In both cases, there is also recuperation of the compound (for *EtOH*, refer Rieger et al. 1983).

Indeed the results can be generalized to attempt to draw wider conclusions.

Of the general balance of the biphasic chemostat in steady state (Eq. 3.3.10a), it is easily derived that

$$X^c = \frac{\Phi_{S,m}^0(c) - D\tilde{C}_S^c}{q_S^c} \tag{4.5}$$

where S represents the limiting substrate.

The biomass then appears as the ratio of a flux and a specific rate. The flux (numerator) is none other than the total of substrate present in the cellular phase of the chemostat (namely the net interphasic exchange flux less the hydraulic outflow of the substrate associated with the cells). The specific rate (denominator) is here a measure of the rate at which the substrate is used.

For an increase in the specific rate of use, the readjusted balance is written, at constant flow as:

$$X^{/c} = \frac{\Phi_{S,m}^0(c) - D\tilde{C}_S^c}{q_S^c + \delta q_S^c} \tag{4.6}$$

where δq_S^c is a nonnegative, finite value.

"At constant flux" meaning that the numerators are equal, it is easily obtained by dividing (Eq. 3.6) by (Eq. 3.5) that

$$\frac{X^{/c}}{X^c} = \frac{q_S^c}{q_S^c + \delta q_S^c} \leq 1 \tag{4.7}$$

or even that,

$$\frac{X^{/c}}{X^c} = \frac{1}{1+\theta} \tag{4.8}$$

where it has been written that,

$$\theta = \frac{\delta q_S^c}{q_S^c} \tag{4.9}$$

This last value is nothing other than the relative value of the specific metabolization rate disturbance (in the broad sense).

The interpretation of these results (Eqs. 4.5–4.7) is then as follows.

If the intracellular flux of substrate is constant, all increase in intracellular consumption rate leads to a reduction in biomass.

A more intuitive (but less rigorous) way makes it possible to better understand this conclusion. Let us write the general balance in another form,

$$\Phi_{S,m}^c(c) - D\tilde{C}_S^c = q_S^c X^c \tag{4.10}$$

The constancy of the substrate flux is expressed by

$$\Phi_{S,m}^c(c) - D\tilde{C}_S^c = \text{ cte.} \tag{4.11}$$

and so

$$q_S^c X^c = \text{ cte.} \tag{4.12}$$

In these circumstances, it is clear that if q_S^c increases, X^c diminishes.

The significance of this phenomenon is that if the specific metabolization rate increases, a smaller quantity of biomass is necessary to handle the same amount of substrate (at constant substrate flux).

In the case of energy dissipation, it is shown that the total specific consumption rate increases due to recirculation (cf. Eq. 3.4.21), whereas in the respirofermentative change, for the sake of simplicity, we will say that the specific rate suddenly increases at a given critical dilution rate that corresponds to the activation of the low affinity metabolic pathway (refer to Fig. 3.21).

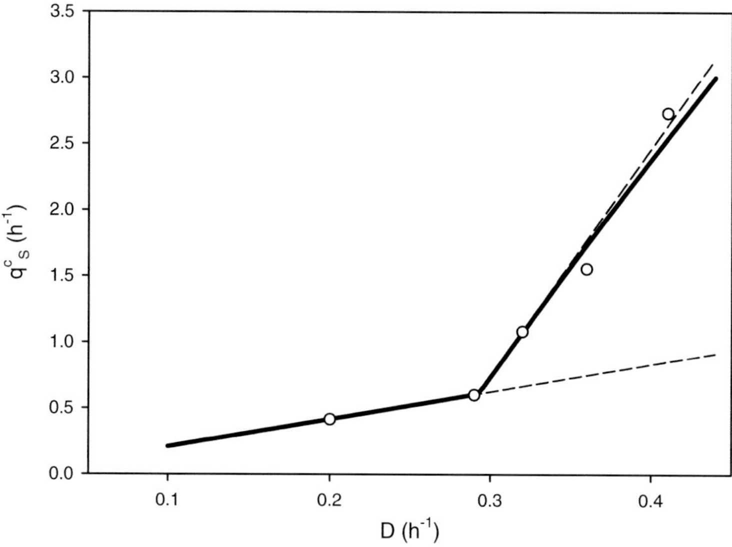

Fig. 4.3 *SIER and specific substrate consumption rate use depending on D*. The *SIER* (as defined by Eq. 3.6.41) is represented by *dots* over the whole of the domain *D*. q_S^c is represented as a *solid line*. The *open circles* are the experimental values (for $S^0 = 5$ g *GLU*/L; refer to Sect. 3.6; the conditions are the same as in Fig. 3.28)

In the case of the Crabtree Effect, it is remarkable to observe (Fig. 3.28) how independent the specific exchange rate is of the influent substrate concentration (similarly moreover for the yield coefficient; refer to Fig. 3.29).

In Fig. 4.3, the *SIER* (dotted line) and the substrate consumption rate (solid line) are represented as a function of *D*. It is noted that the two curves are overlap exactly for $D < D_c$ (≈ 0.3 h^{-1}) and diverge very little at high dilution rates. By dividing the balance (Eq. 4.10) on the left-hand and the right-hand sides by X^c and by using the definition of the yield coefficient, it is shown that

$$q_S^c \approx \frac{D}{Y_{X^c,S}} \tag{4.13}$$

over the whole of the domain. *D/Y* is the maximum value that can take the specific rate of use of substrate (refer to case D, in Sect. 3.3, when (Eq. 3.3.23) is satisfied).

Equation (4.13) also implies that $\tilde{C}_S^c \approx 0$, that is to say that the intracellular concentration of substrate is about nil. (The theoretical Figs. 3.18 and 3.19 show the accumulation of cell substrate and the difference between *D/Y* and q_S^c respectively.) However, the difference between the rate of transfer and the maximum rate of use increases with *D* and is maximum at the washout. Moreover, for D ≈ 0.3 h^{-1}, the yield coefficient diminishes, thereby showing that the global capacities of biosynthesis are exceeded.

The flux of transfer constancy, the adjustment of mass balances via biomass, the attempt to exceed the maximum value of *SIER* (refer to Eq. 3.4.28) the (possible?) washout due to exceeding the processing capacity and the accumulation of intracellular substrates, all this offers the image of a cell physiology that tries to maximize the rate of transfer of the external medium towards the intracellular medium. The substrate seems primarily affected to biomass production, but if the transfer rate exceeds the biosynthesis (or more generally the primary metabolism) diverse strategies seem able to set up, such as substrate expulsion with energy dissipation, or activation of a side fermentation pathway including secondary metabolites excretion. This last observation is confirmed by Ostergaard et al. (2000) who, using genetic eliminated metabolic inhibitors of galactose in *Saccharomyces cerevisiae*. As a result, there was increased substrate consumption (with a flux increase of more than 40 % in batch). The final result was not an increase in biomass, but an increase in respirofermentative metabolism, with improved production of ethanol, proportional to the galactose flux. Rossel et al. (2002) have also shown a correlation between the transport of glucose and the fermentation capacity of the commercial baker's yeast and they emphasize how little attention has been given to substrate transport phenomena in the study of experiences of this nature. Nevertheless, maximization of substrate transport rate can at this stage, only be considered as a working hypothesis and not as a general principle.

In conclusion, our objectives were to construct a method of representation (explicit or implicit) adapted to polyphasic systems where the dispersion of phases (PDS) makes possible a pseudo-homogenous approach and different level of description.

The general theoretical aspect was applied to the biphasic chemostat, a system that is both simple in conception and well-illustrated in the literature. The goal, this time was to show that the PDS approach could be useful, possess sufficient originality to add to the analysis of certain phenomena and be sufficiently heuristic to initiate new lines of thought. It is thought that this has been shown and the objectives have been achieved.

Regrets have been expressed because it was not possible to apply the theory to other systems than the biphasic chemostat. The discussion about applications shows well how far it has been possible to go in the biological interpretation of examples chosen as illustrations of the theory.

This work is considered as the beginnings of a step forward and not as an ending. The authors greatest desire is one day there will be either for the short or long term, a continuation of what has been initiated here that is independent of form, means, and circumstances.

References

Blumenfeld L.A. and Tikhonov A.N. (1994) Biophysical thermodynamics of intracellular processes. Springer-Verlag New York Inc.

Gardner R.C. (2003) Genes for magnesium transport. Current opinion in plant biology. **6**:263–267.

Mandelstam J., McQuillen K. and Dawes I. (1982) Biochemistry of bacterial growth. Blackwell Scientific Publications (Third Edition).

Ostergaard S., Olsson L., Johnston M. and J. Nielsen. (2000) Increasing galactose consumption by *Saccharomuces cerevisiae* through metabolic engineering of the *GAL* gene regulatory network. Nature Biotechnology **18:** 1283–1286.

Rieger M., Käpelli O. and Fiechter A. (1983) The role of limited respiration in the complete oxidation of glucose by *Saccharomyces cerevisiae*. J. Gen. Microbiol. **129:** 653–661.

Rossel S., van der Weiden C.C., Kruckeberg A., Bakker B.M. and H.V. Westerhoff (2002) Loss of fermentative capacity in baker's yeast can partly be explained by reduced glucose uptake capacity. Molecular Biology Reports. **29**:255–257.

Smith R.L., Banks J.L. Snavely M.D. and Maguire M.E. (1993) Sequence and topology of the CorA magnesium transport systems of *Salmonella typhimurium* and *Escherichia coli* – identification of a new class of transport protein. J. Biol. Chem. **268**: 14071–14080.

Smith R.L. and Maguire M.E. (1995) Distribution of the CorA Mg^{2+} transport system in gram-negative bacteria. J. Bacteriol. **177:** 1638–1640.

Tempest D.W., Hunter J.R. and Sykes J. (1965) Magnesium limited growth of *Aerobacter aerogenes* in a chemostat J. Gen. Microbiol. 39: 355–366.

Appendix A.1
Theory

Foreword:
The theory that appears in this appendix is more complete than that in the body of the text. This is because attempts were made to develop the most coherent and general theory possible, that included the greatest number of cases possible. Obviously, not all these cases have been treated, but it seemed useful to keep the most general in view, of possible future applications. It is clear that the representation as a PDS reduces the number of levels of description in a big way; structured representation (that takes into account each micelle in the system rather than an average micelle) again increases this number of levels. The designation of each quantity becomes more and more complex when the number of phases increases (and if the system is structured). The perfecting of a coherent and theory, therefore becomes an arduous task, but an indispensable one if complex systems are to be developed.

A.1.1 Levels of Description

There are three levels of description, denoted as three types:

- micellian level (or of the micelle): TYPE 1
- phase level (or of the phase): TYPE 2
- system level (or of the system): TYPE 3.

A.1.2 Unstructured Variables

The unstructured variables do not apply to PDS for which it is possible to define average quantities (mean size micelle, intramicellian mean concentrations, mean cell age, etc. (cf. Chap. 2)).

By convention, extensive variables that are applied to the micelle level (type 1) are represented by lowercase letters. In an unstructured representation, quantities of

J. Thierie, *Introduction to Polyphasic Dispersed Systems Theory*,
DOI 10.1007/978-3-319-27853-7

type 1 overlap with intensive quantities of type 2 (average phase value being identical to the average value of the micelles). All other quantities are represented by uppercase letters.

The units are indicated between square brackets [...], with,

V = volume
M = mass
T = time
* = none (dimensionless numbers)

T indicates the variable type (refer to Sect. A.1.1)
D is the definition number (given in Appendix A.2)

Latin characters

$\tilde{C}_i$	Pseudo-homogeneous mass concentration of the compound i in the system (P-concentration)	[M V^{-1}]	T3	D1
C_i^p	Mass concentration of the compound i in the micelle of phase p (R-concentration)	[M V^{-1}]	T1	D2
$\tilde{C}_i^p$	Partial pseudo-homogeneous mass concentration of the compound i in phase p (E-concentrations)	[M V^{-1}]	T2	D3
$C_i^E(n)$	Mass concentration of compound i at the inlet n of the system (*)	[M V^{-1}]	T3	
$C_i^S(n)$	Mass concentration of the compound i at the outlet n of the system (*)	[M V^{-1}]	T3	
$\tilde{C}_i^{p,E}(n)$	Partial pseudo-homogeneous mass concentration of the compound i at the inlet n of the system (*)	[M V^{-1}]	T3	
$\tilde{C}_i^{p,S}(n)$	Partial pseudo-homogeneous mass concentration of the compound i at the outlet n of the system (*)	[M V^{-1}]	T3	
$F_i^E(n)$	Mass flow of the compound i at the inlet n (*)	[M t^{-1}]	T3	D4
$F_i^S(n)$	mass flow of the compound i at the outlet n (*)	[M t^{-1}]	T3	D4
$\tilde{F}_i^{p,E}(n)$	Partial pseudo-homogeneous mass flux of the compound from phase p at the inlet n of the system (*)	[M t^{-1}]	T3	
$\tilde{F}_i^{p,S}(n)$	Partial pseudo-homogeneous mass flow of the compound i from phase p at the outlet n of the system (*)	[M t^{-1}]	T3	
FE_i^p	Sum of the net exchange flows of the compound i across N_{p-1} interfaces, related to phase p by unit of working volume	[M (V t)$^{-1}$]	T2	D5
M_i	Total mass of i in the system	[M]	T3	
M^p	Mass of phase p	[M]	T2	
M_i^p	Mass of compound i in phase p	[M]	T2	
m^p	Average mass of a micelle from phase p	[M]	T1	
m_i^p	Average mass of compound i in the phase p micelle	[M]	T1	
N_p	Number of different phases	[*]	T3	
N_T^p	Number of micelles in phase p	[*]	T2	

(continued)

NC^p	Number of different compounds in phase p	[*]	T2	
NF^p	Number of different families of compounds in phase p	[*]	T2	
$Q^E(n)$	Global volume flow at the inlet n(*)	[V t^{-1}]	T3	
$Q^S(n)$	Global volume flow at the outlet n(*)	[V t^{-1}]	T3	
$r_i^p(.)$	Kinetics of the compound i in the phase p micelle	[M V^{-1} t^{-1}]	T1	
V^p	Volume of phase p	[V]	T2	
V_T	Total working volume of the system	[V]	T3	D6
v^p	Mean volume of a micelle from phase p	[V]	T1	
X^p	Phase density of phase p	[M V^{-1}]	T3	D14

(*) There is only one inlet (outlet) for these variables, therefore the expression between parentheses is omitted

Greek Characters

α_i^p	Mass fraction of the compound i in the micelle (or in the phase)	[M M^{-1}]	T1 (2)	D7
δ_p	Mean specific mass of phase p or in the micelle (density)	[M V^{-1}]	T2	D8
Δ_p	Micellian density	[* M^{-1}]	T3	D9
$\Phi_{i,\{q\neq p\}}(p)$	Net exchange flux of the compound i across all interfaces $\{q \neq p\}$ of phase p	[M t^{-1}]	T2	D10
φ_i^p	Mass distribution coefficient i	[M M^{-1}]	T3	D11
$\phi_{i,p}^E(q)$	Interfacial mass inflow of compound i in the micelle of phase p across the interface p, q	[M t^{-1}]	T1	D12
$\phi_{i,p}^S(q)$	Interfacial mass outflow of compound i in the micelle of phase p across the interface p, q	[M t^{-1}]	T1	D12

A.1.3 Structured Variables—General Principles

In a structured theory, the notation of the quantities of type 1 must make it possible to distinguish one micelle from another. Just a few approaches are given here to construct a complete and coherent theory.

A.1.3.1 Masses and Concentrations

$m_{i,j}^p$ mass of compound i in micelle j from phase p

$c_{i,j}^p$ concentration of compound i in micelle j of phase p.

A.1.3.2 General Intermicellian Flow

$\phi_{i,p,k}^{E/S}(q,l)$ mass flux at inlet E (or at outlet, S) of the compound i in the micelle k of phase p across the interface with micelle l of phase q (related to phase p).

Appendix A.2
Definitions

D1. **P-concentrations**: Pseudo-homogeneous Mass Concentration.

$$\tilde{C}_i = \frac{M_i}{V_T}$$

Ratio of the total mass of the compound i to the total working volume of the system. The compound i can be divided in a non-homogeneous way between the different phases, this concentration is a 'virtual' quantity that does not have a defined biochemical meaning. In theory, it corresponds to the definition of mass concentration of a homogeneous system (where it has a precise meaning), hence its name *pseudo-homogeneous*.

D2. **R-concentrations**.

$$C_i^p = \frac{m_i^p}{v^p} \tag{D2.a}$$

Basically, it is a type 1 quantity, relevant for intra-micellian kinetic description (hence the name R-concentration or R̲eacting concentration), and corresponds to the ratio of compound i in the micelle and the volume of the micelle. However, in a non-structured representation, these quantities are average values, and so equal in all micelles. There is also,

$$C_i^p = \frac{N_T^p . m_i^p}{N_T^p . v^p} = \frac{M_i^p}{V^p} \tag{D2.b}$$

This second definition resembles a type 2 quantity. It must only however, be taken as the *measurement* of mean R-concentration and not as its definition. To clarify this, let's consider a *structured* system. So there is,

$$C_{i,j}^p = \frac{m_{i,j}^p}{v_j^p}$$

where j denotes a micelle that is well defined (the jth micelle).

J. Thierie, *Introduction to Polyphasic Dispersed Systems Theory*,
DOI 10.1007/978-3-319-27853-7

It is very much on this quantity that the kinetics inside the micelle j depend. This quantity has a well defined (bio) chemical meaning. However, there is nothing to stop this being written,

$$C_i^p = \sum_j \frac{m_{i,j}^p}{v_j^p} = \frac{M_i^p}{V^p}$$

This value is in all ways identical to (D2.b) but no case whatsoever can it be used to describe the intra-micellian kinetics in a structured system.

***D3*. E-concentration**: Partial Pseudo-homogeneous Mass Concentration.

$$\tilde{C}_i^p = \frac{M_i^p}{V_T}$$

The ratio between the mass of the compound i throughout phase p and the total working volume of the system. Just as for the P-concentration, this quantity does not generally have a precise (bio) chemical sense but is a convenient definition. The ratio of a type 2 and a type 3 quantity is a quantity of a type that is not well defined. However, it is defined by preference as type 2, having regard to the fact that it concerns the mass that is spread throughout phase p.

***D4*. Input/Output Massic Fluxes**

These quantities characterize the inlets and outlets of the system in general. They can characterize liquid, gaseous and even solid fluxes. A general definition of these fluxes is therefore impossible.

Liquid Flows

The liquid massic fluxes are crucial; they are defined by,

$$F_i^A(n) = Q^A(n) \times C_i^A(n)$$

A product of the volumetric flux and the pseudo-homogeneous (or homogeneous) concentration that corresponds to the exchange point n of the system with the external environment (A corresponding to an inlet or an outlet). Generally, it is not necessary to distinguish the different inlet or outlets and the index n can be omitted, the flow then representing, the sum of the inflows or outflows.

By definition,

$$F_i^A(n) \geq 0; \quad \forall\ A, n$$

The sign corresponding to the inlet (+) or to the outlet (−) intervening only in the mass balance.

D5. Sum of Net Exchange Fluxes

$$FE_i^p = \frac{\Phi_{i,\{p \neq q\}}(p)}{V_T}$$

The sum of the net exchange fluxes per volume unit represents the inputs/outputs balance of the compound i in the core of phase p, across all the interfaces of all the micelles belonging to this phase. FE_i^p is the general introductory form redefined afterwards by $\Phi_{i,\{p \neq q\}}(p)$.

D6. Total System Working Volume

The total volume of the system is a less obvious notion that is appearing. Basically, *working* signifies required for the description of the phenomenon. The volumes of inert, submerged objects in the sense that they do not interfere with the studied process (sensors, agitators, diffusers, etc.) should not be taken into account for the calculation of working volume. On the other hand, certain reactions can be produced outside the body of the reactor (in the inlet and outlet pipes of internal recycling, for example). If this effect is not insignificant, it should be taken into account.

The number of phase count is even more delicate. It can be agreed that one of the phases is not useful for the description of a process (for example the gaseous phase in a triphasic system (solid, liquid, gas) because only the dissolved gases (that is to say in liquid phase) are relevant). Therefore, it is legitimate to exclude the volume of this phase from the working volume. On the other hand, this would not be the case if this phase was included to follow the evolution of gaseous and/or volatile compounds. The necessary condition of checking that the working volume corresponds to the sum of the volumes of the phases studied (if not, the accuracy of the mass balance cannot be guaranteed).

D7. Mass Fraction

$$\alpha_i^p = \frac{m_i^p}{m^p} \tag{D7.a}$$

The ratio of the mass of the compound i in the micelle and the total mass of the micelle. In the unstructured representation, the micellian mass fraction is equal to the phase mass fraction,

$$\alpha_i^p = \frac{N_T^p.m_i^p}{N_T^p.m^p} = \frac{M_i^p}{M^p} \tag{D7.b}$$

This last relationship is, however a measurement of the mass fraction and not its definition (refer to D2 and the comment concerning the R-concentrations).

D8. Specific Mass

$$\delta_p = \frac{m^p}{v^p} \tag{D8.a}$$

Ratio of micelle mass to its volume. In an unstructured representation the specific micelle mass is equal to the specific mass of the phase.

$$\delta_p = \frac{M^p}{V^p} \tag{D8.b}$$

Just as in *D2* or *D7*, this last relationship is a measure of the specific mass and not a definition.

D9. Micellian Density

$$\Delta_p = \frac{N_T^p}{V_T}$$

Ratio of the number of micelles in phase p to the total working volume. It is a quantity of mixed type (types 2 and 3) that acquires a meaning at the system level and that is classed as type 3 because of this.

D10. Net Phasic Exchange Flow

$$\Phi_{i,\{q \neq p\}}(p) = N_T^p \cdot \sum_{q \neq p}^{N_p} \left(\phi_{i,q}^E(p) - \phi_{i,q}^S(p) \right) \tag{D10.a}$$

Sum over all micelles in phase p of the difference between micellian inlet and the outlet fluxes across the interfaces p and $\{q \neq p\}$. The quantity concerns, therefore all interfaces of the same nature, whether between p and. It is a type 2 quantity, but the definition above is only true in an unstructured representation. In a structured representation, this must be written,

$$\Phi_{i,\{q \neq p\}}(p) = \sum_{k=1}^{N_T^p} \sum_{q \neq p}^{N_p} \left(\phi_{i,q}^E(p) - \phi_{i,q}^S(p) \right) \tag{D10.b}$$

where k is the index of the kth micelle of phase p.

The (bio) chemical sense is obviously relevant and identical in both cases.

The sign of the net exchange flow is not defined a priori and corresponds globally to an input in the phase if $\Phi_{i,\{q \neq p\}}(p) > 0$ and to an output if $\Phi_{i,\{q \neq p\}}(p) < 0$.

D11. Massic Distribution Coefficient

$$\varphi_i^p = \frac{M_i^p}{M_i}$$

Ratio of mass of compound i in phase p and the mass of compound i throughout the system. This dimensionless quantity measures the degree of distribution I between the different phases. It has the same physical meaning as the partition coefficient in polyphasic undispersed systems. A quantity of mixed type (type 2 and 3), it takes its meaning to system level and is therefore classed as type 3.

D12. Interfacial Micellian Mass Flux

$\phi_{i,p}^A(q)$ is the mass of compound i per unit of time that crosses the surface of the interface p, q of the micelle of phase p ($A = E$: input; $A = S$: output).

By definition, $\phi_{i,p}^A(q) \geq 0;\ \forall A$. The sign of this flow must be positive for an inflow in phase p and negative for the outflow of this, if phase p is chosen as a standard. It is clear, however, that the quantity of matter than flows in or out in a micelle must be equal to the quantity of matter that leaves or enters by exchange with another. In the referential p therefore,

$$\phi_{i,p}^A(q) = -\phi_{i,q}^A(p);\quad \forall\ \ p, q\,(p \neq q)$$

D13. Partial Mass Flux

Liquid Flux

$$\tilde{F}_i^A(n) = Q^A \tilde{C}_i^{p,A}$$

A product of volume flow and partial pseudo-homogeneous concentration that corresponds to the exchange point n of the system with the external environment refer to ($A = E$: input; $A = S$: output). For the definition of the sign, refer to D4.

D14. Phase Density

$$X^P = \frac{M^p}{V_T}$$

Ratio of the total mass of phase p to the total working volume.

When phase p is a cell phase, the density takes a particular meaning, namely that it represents biomass (from phase p; it can have several cell phases, for example, in the case of consortiums). To preserve the usual notation for biomass, the symbol X is used for the phase density.

Appendix A.3
Specific Growth Rate for the Whole

Let's look again at the balance evolution equation for a metabolite i of the cell phase c in the form that has already been used in Sect. 3.3.

$$\frac{d\tilde{C}_i^c}{dt} = -D\tilde{C}_i^c + \sigma_{qi}\left|q_i^c X^c\right| + \sigma_{\Phi i}\left|\Phi_{i,x}^0(y)\right| + \tilde{C}_i^c \frac{d \ln N_T^c}{dt} \tag{A.3.1}$$

where σ_{wz} is the sign of the quantity wz (-1, 0, 1) and $\Phi_{i,x}^0(y)$ the exchange flow between the phases per unit of working volume (x, y can take only the values c and m here; the direction of the flow is in the sense of $x \to y$). The conservation of matter obviously implies that (refer to 3.3.9):

$$\Phi_{i,c}^0(m) = -\Phi_{i,m}^0(c) \tag{A.3.2}$$

A more compact form of (A.3.1) is obtained by putting

$$\lambda_i^c = \sigma_{qi} q_i^c + \sigma_{\varphi i}\Phi_{i,x}^0(y) \tag{A.3.3}$$

and so

$$\frac{d\tilde{C}_i^c}{dt} = -D\tilde{C}_i^c + \lambda_i^c + \tilde{C}_i^c \frac{d \ln N_T^c}{dt} \tag{A.3.4}$$

taking that into account:

$$\tilde{C}_i^c = \alpha_i^c X^c$$

where α_i^c is the mass fraction of the compound i (ratio of the mass of this compound to the total mass of the phase $\alpha_i^c = M_i^c / M^c$), Eq. (A.3.4) takes the form,

$$\frac{d\tilde{C}_i^c}{dt} = -\alpha_i^c D X^c + \lambda_i^c + \alpha_i^c X^c \frac{d \ln N_T^c}{dt} \tag{A.3.5}$$

J. Thierie, *Introduction to Polyphasic Dispersed Systems Theory*,
DOI 10.1007/978-3-319-27853-7

The term λ_i^c is the only one that relates to the modifications in the micelles (cells) of the phase c; it is called the partial endogenous flow (partial, because it relates to the compound i). By definition, the biomass is the sum of all the compounds that make up the cell phase (including water)

$$X^c = \sum_{i=1}^{NC^c} \tilde{C}_i^c$$

where NC^c is the total number of different compounds in phase c. Then the law of evolution of the biomass is obtained by applying the principle of grouping (refer to Sect. 2.5) to all the compounds of the phase. By applying this principle to the Eq. (A.3.5) and taking into account that, then,

$$\frac{\mathrm{d}X^c}{\mathrm{d}t} = -DX^c + \sum_{i=1}^{NC^c} \lambda_i^c + X^c \frac{\mathrm{d}\ln N_T^c}{\mathrm{d}t}$$

and by defining the total endogenous flow by

$$\lambda^c = \sum_{i=1}^{NC^c} \lambda_i^c$$

it is finally obtained that:

$$\frac{\mathrm{d}X^c}{\mathrm{d}t} = -DX^c + \lambda^c + X^c \frac{\mathrm{d}\ln N_T^c}{\mathrm{d}t} \tag{A.3.6}$$

This equation is very important because it shows in a rigorous way the link between the exchange flux linking the external world, the total endogenous flux and the variation in the number of cells.

Equation (A.3.6) can be given a more usual form by putting

$$\mu^c \doteq \frac{\lambda^c}{X^c} \tag{A.3.7}$$

where μ^c has the dimensions of a specific speed. The relationship (A.3.6) is then written

$$\frac{\mathrm{d}X^c}{\mathrm{d}t} = X^c \left(\mu^c - D + \frac{\mathrm{d}\ln N_T^c}{\mathrm{d}t} \right) \tag{A.3.8}$$

In the steady state, the number of cells is constant and the condition of stationarity of (A.3.8) is written

$$\mu^c = D \tag{A.3.9}$$

Let's come back to Eq. (A.3.6): by expliciting λ^c, it takes the form:

$$\frac{1}{X^c}\frac{\mathrm{d}X^c}{\mathrm{d}t} = -D + \frac{1}{X^c}\sum_{i=1}^{NC^c}\left(\sigma_{qi}q_i^c + \sigma_{\Phi i}\Phi_{i,x}^0(y)\right) + \frac{\mathrm{d}\ln N_T^c}{\mathrm{d}t} \tag{A.3.10}$$

This form is also very interesting, because it shows that the specific growth rate [the right hand side of (A.3.10)] is the result (the sum) of the contribution of all the processes of anabolism/catabolism of the metabolites (via q_i^c) and the exchange processes between phases (via Φ_i^0). So it can be expected that this specific rate is a more varied behavior in a culture supplied by complex substrate compared with a culture growing on a simple medium. Notably, because the number of physiological mechanisms and exchanges brought into play in the complex case should be superior to that of simple substrate. It is in fact what is observed.

Figure A.1 shows the differences in behavior in the production of biomass in batch (D = 0). In the case of a complex substrate, the growth curve is significantly distanced from a simple exponential. To take up again Herbert's terms

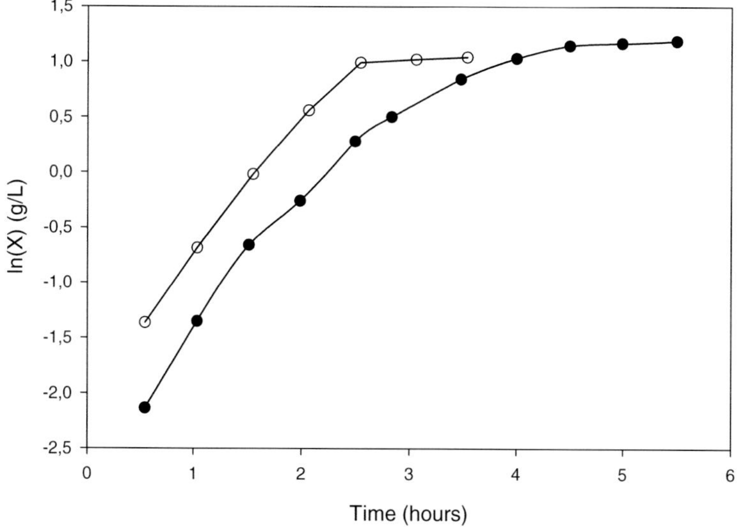

Fig. A.1 Pseudo-stationary state approach in batch. Profiles of the approach to the pseudo-stationary state in a stirred fermentor where *Enterobacter* (*Aerobacter*) *aerogenes* is cultivated in a complex medium (*filled circle*) and in a simple medium (*open circle*) (According to Strange et al. (1961), cited in Herbert 1961)

(Herbert 1961): "*As others have observed in such media (Monod 1949) there are apparently two or perhaps three phases of exponential growth at successively decreasing rates, probably indicating successive exhaustion of non essential but growth-accelerating nutrients.*" According to our point of view, there are no nutrients that accelerate growth as such. Quite simply, exhaustion of nutrients or reduction in exchange flux, causes certain terms of the sum that figure in the right hand side of the Eq. (A.3.10) to vanish, which leads to a reduction in specific growth rate even if all the other terms remain constant. The phenomenon is particularly clear on approaching a pseudo-stationary state (cf. Fig. A.1): in the simple medium, growth stops suddenly whereas in the complex medium, growth ending is progressive, showing the contribution of several factors to the establishment of the resulting rate. In other terms, in the simple medium, the relationship Eq. (A.3.10) is probably reduced to a small number of prevailing terms (perhaps even just one), governed by the principal substrate, whereas in the complex medium, a greater number of terms, with comparable weighting contribute to the resulting rate.

Comment: Variational term appearing in (A.3.8) may seem 'amazing'. Appendix A.5 explains the interpretation to give to this term that is not properly speaking a cell production in the same sense as growth.

References

Herbert D. (1961) The chemical composition of microorganisms as a function of their environment. 11th Symp. Soc. Gen. Microbiol. **11**: 391–416. Cambridge University Press.

Monod J. (1949) The growth of bacterial cultures. Ann. Rev. Microbiol. **3** : 371–394.

Strange, R. E., Dark, F. A. & Ness, A. G. (1961). The survival of stationary phase cells of *Aerobacter aerogenes* stored in aqueous suspension. J. gen. Microbiol. **25**: 61–76.

Appendix A.4
Cancelling of Interphasic Exchange Fluxes at the System Level

One of the interesting characteristics of PDS is that, when a partition of the system into its different phases occurs, interphasic flux are highlighted that are cryptic in other methods of representation. Much information can be drawn from these fluxes. Conversely, when it comes to calculating the general balance by addition of micelles and phases, the sum over the interphasic fluxes is cancelled.

So it is to be shown that

$$\sum_{p=1}^{N_p} \Phi_{i,\{q \neq p\}}(p) = 0 \tag{A.4.1}$$

(Refer to Sect. 2.4. concerning the balances for the signification of the symbols.). By using (2.4.4) and (2.4.8), (A.4.1) can be put in the form,

$$\sum_{p=1}^{N_p} N_T^p \sum_{q \neq p}^{N_p} \left(\phi_{i,q}^E(p) - \phi_{i,q}^S(p) \right) = 0 \tag{A.4.2}$$

By rearranging,

$$\sum_{p=1}^{N_p} \sum_{q \neq p}^{N_p} N_T^p \left(\phi_{i,q}^E(p) - \phi_{i,q}^S(p) \right) = 0 \tag{A.4.3}$$

For simplicity of the formalism, put,

$$\Delta_{i,q}(p) = N_T^p \left(\phi_{i,q}^E(p) - \phi_{i,q}^S(p) \right)$$

This quantity expresses the flux leaving the phase q to enter into phase the p. The relationship (A.4.2) then takes the form:

J. Thierie, *Introduction to Polyphasic Dispersed Systems Theory*,
DOI 10.1007/978-3-319-27853-7

$$\sum_{p=1}^{N_p}\sum_{q\neq p}^{N_p}\Delta^p_{i,q}(p)=0 \tag{A.4.4}$$

The flux of matter that leaves one phase for another is obviously equal to the quantity of matter that enters into the arrival phase, so that conservation of matter implies that

$$\Delta_{i,q}(p)=-\Delta_{i,p}(q) \tag{A.4.5}$$

The double summation (A.4.4) is a sum of $N_p(N_p-1)$ terms ($N_p(N_p-1)$ is an even value, since the product of one an even integer with any other integer is an even value; now if N_p is odd, N_p-1 is even and vice-versa).

The accuracy of (A.4.1) is therefore established if it is shown that in (A.4.4), for all $\Delta_{i,k}(l)$ terms there is a corresponding $\Delta_{i,l}(k);\forall\, l,k$ term.

Let's fix $p = k$ in the first summation of (A.4.4); given that the second summation is done over all the values of q (except k), a $\Delta_{i,k}(l)$term necessarily appears. Let's fix now $p = l$ in the first summation; then a $\Delta_{i,l}(k)$ term necessarily appears. In this way it comes about that the relationship Eq. (A.4.4) is formed of $N_p(N_p-1)$ terms forming $N_p(N_p-1)/2$ couples $\{\Delta_{i,k}(l),\Delta_{i,l}(k)\}$ that, in accordance with the mass balance Eq. (A.4.5), are cancelled two by two. So it can be shown that the sum of all phases of interphasic fluxes is cancelled.

Appendix A.5
Consideration about the Micellian Variational Term

The micellian variational term is a term that naturally appears as soon as it is considered that the number of micelles is not constant (cf. 3.4.10). However, it is not, by construction, an input/output term, but rather a combinatorial term that takes into account the distribution of an intramicellian compound when such or such a micelle appears or disappears in the concerned phase. In this appendix attempts are made to clarify the concept a little.

If a homogeneous closed system (classic) of volume V_T is considered, the mass balance equation for any compound i is written

$$\frac{\mathrm{d}M_i}{\mathrm{d}t} = R_i(.)V_T \tag{A.5.1}$$

where $R_i(.)$ is the whole of the transformations reactions of the compound within the system.

Now let's take the balance (2.4.12) and lets sum up all the phases; so

$$\frac{\mathrm{d}M_i}{\mathrm{d}t} = \sum_p r_i^p(.)V^p + \sum_p M_i^p \frac{\mathrm{d}\ln N_T^p}{\mathrm{d}t} \tag{A.5.2}$$

(the interphasic exchange terms are cancelled; refer to Appendix A.4).

The two representations can be applied to the same system, because the left hand side of (A.5.1) and (A.5.2) only express the global variation of i inside the system, independently of the modalities of the processes that are taking place (homogeneous or heterogeneous). If the intramicellian kinetics are equivalent ($R_i(.)$ representing the global kinetic and $\sum_p r_i^p(.)$ the sum of the constituents in the different phases) it should be conceded that

$$R_i(.)V_T \equiv \sum_p r_i^p(.)V^p \tag{A.5.3}$$

J. Thierie, *Introduction to Polyphasic Dispersed Systems Theory*,
DOI 10.1007/978-3-319-27853-7

Equations (A.5.1) and (A.5.2) are then two equivalent representations of the same system.

This obviously implies (comparing (A.5.1) and (A.5.2) under condition (A.5.3)) that,

$$\sum_p M_i^p \frac{\mathrm{d} \ln N_T^p}{\mathrm{d}t} = 0 \tag{A.5.4}$$

This condition is evidently satisfied if the number of micelles is constant in all phases, but this is not the general case.

The general case (where the number of micelles varies) must then be interpreted as a series of change in the number of micelles that compensate each other. Each increase in the number of micelles in a phase must be compensated for by a corresponding reduction in one of several other phases, the variations being weighted by a factor that depends on the quantity i in each phase. To make this point clear, let's re-write (A.5.4) in the following form,

$$M_i \sum_p \varepsilon_i \frac{\mathrm{d} \ln N_T^p}{\mathrm{d}t} = 0 \tag{A.5.5}$$

where

$$\varepsilon_i = \frac{M_i^p}{M_i}; \quad 0 \leq \varepsilon_i \leq 1 \tag{A.5.6}$$

This coefficient represents the fraction of i in the phase p with respect to the total quantity of i in the system. For $M_i \neq 0$, the condition (A.5.5) is written,

$$\sum_p \varepsilon_i \frac{\mathrm{d} \ln N_T^p}{\mathrm{d}t} = 0 \tag{A.5.7}$$

Each variational term of (A.5.7) is then weighted by a coefficient that depends on the quantity relative to i in the corresponding phase.

Discussion: The variational term that appears on the right hand side of (2.4.12) indicates that the phase p is open in with respect to the number of micelles (some can be created, other disappear). On the other hand, this variational term does not imply in any way whatsoever that there is an opening at the entire system level. In reality (2.4.12) represents the mass balance in an open phase embedded in a closed system as we have shown previously by the equivalence of closed systems (A.5.1) and (A.5.2).

In a dense (without dispersing phase) and closed system, all appearances (or disappearances) of a micelle from phase p must necessarily be accompanied by a transfer of the compound i from (or towards) one or several micelles. In these conditions, the relationship (A.5.4) or (A.5.7) represents a combinatorial term, a way of representing how takes place the distribution of the compound i between the different phases when the number of micelles varies at within of the system. Dense, diluted or compact systems will have a qualitatively different behaviour from this point of view, but this does not detract from the fact that the variational terms from (2.4.12) do not imply in any way whatsoever an opening of the system. This last observation leads us to the introduce the input/output terms in (2.4.13).

Printed in the United States
By Bookmasters